Ulrich Lange

Electrochemical Transistor and Chemoresistor based Sensors:

Ulrich Lange

Electrochemical Transistor and Chemoresistor based Sensors:

Measurement Technique, Materials and Applications

Südwestdeutscher Verlag für Hochschulschriften

Imprint
Any brand names and product names mentioned in this book are subject to trademark, brand or patent protection and are trademarks or registered trademarks of their respective holders. The use of brand names, product names, common names, trade names, product descriptions etc. even without a particular marking in this work is in no way to be construed to mean that such names may be regarded as unrestricted in respect of trademark and brand protection legislation and could thus be used by anyone.

Publisher:
Südwestdeutscher Verlag für Hochschulschriften
is a trademark of
Dodo Books Indian Ocean Ltd., member of the OmniScriptum S.R.L Publishing group
str. A.Russo 15, of. 61, Chisinau-2068, Republic of Moldova Europe
Printed at: see last page
ISBN: 978-3-8381-2347-9

Zugl. / Approved by: Regensburg, Universität, Diss., 2010

Copyright © Ulrich Lange
Copyright © 2011 Dodo Books Indian Ocean Ltd., member of the OmniScriptum S.R.L Publishing group

TABLE OF CONTENTS

1. **INTRODUCTION** 1
   1.1 Conducting polymers 1
      1.1.1 Conducting polymer based sensors 4
   1.2 Graphene 5
      1.2.1 Graphene in sensor application 6
   1.3 Metallic nanoparticles 7
      1.3.1 Metallic nanoparticles in sensors 8
   1.4 Conductometric sensors 8
      1.4.1 Chemoresistors 10
      1.4.2 Electrochemical transistors 13
   1.5 Aim of the work 16
   1.6 References 17

2. **METHODS** 23
   2.1 In-situ simultaneous two- and four-point measurement 23
      2.1.1 Theory and working principle 23
      2.1.2 Electrodes 25
   2.2 References 27

3. **RESULTS AND DISCUSSION** 28
   3.1. Simultaneous measurements of bulk and contact resistance 28
      3.1.1. Results and discussion 28
      3.1.2. Experimental 33
      3.1.3. References 33
   3.2. Characterisation of polythiophene in aqueous and organic solutions ... 34
      3.2.1. Results and Discussion 35
      3.2.2. Experimental 40
      3.2.3. References 41
   3.3. Six electrode electrochemical transistor 42
      3.3.1. Six electrode measurements 42
         3.3.1.1. Results and Discussion 42
      3.3.2. Electrochemical regeneration of conducting polymer based gas sensors . 46
         3.3.2.1. Results and Discussion 46
      3.3.3. Experimental 49
      3.3.4. References 50

3.4. Electrochemical transistors with ion selective gate electrodes 52
  3.4.1. Results and Discussion 52
  3.4.2. Experimental 54
  3.4.3. References 55

3.5. Polyaniline metal nanoparticle layer by layer composites 56
  3.5.1. Polyaniline gold nanoparticle composite 56
    3.5.1.1. Results and discussion 57
  3.5.2. Polyaniline palladium nanoparticle composite 64
    3.5.2.1. Results and Discussion 64
  3.5.3. Experimental 72
  3.5.4. References 74

3.6. PEDOT / PSS palladium nanoparticle composite 77
  3.6.1. Results and Discussion 77
  3.6.2. Experimental 85
  3.6.3. References 85

3.7. Graphene based gas sensors 87
  3.7.1. Graphene characterisation 87
    3.7.1.1. Results and Discussion 87
  3.7.2. Evaluation of graphene as sensor material for NO_2 sensing 92
    3.7.2.1. Results and Discussion 92
  3.7.3. Graphene palladium nanoparticle layer by layer composite 94
    3.7.3.1. Results and discussion 95
  3.7.4. Electrochemical modification of graphene with nanoparticles 101
    3.7.4.1. Results and Discussion 101
  3.7.5. Experimental 103
  3.7.6. References 104

4. CONCLUSION 107

1. Introduction

Conducting polymers and carbon nanomaterials like carbon nanotubes and graphene are promising materials for chemical and biological sensors,[1]-[5] due to their ability to work as receptor and transducer in such devices. Chemoresistors based on these materials are up to now mainly used as gas sensors, however can also be used to monitor pH, concentration of redox active species, ion concentrations, protein and DNA interactions and biochemical reactions. If the electrochemical potential of the sensor film is controlled by applying a potential versus a reference electrode the setup is called electrochemical transistor, due to an analogy to field effect transistors. A detailed description of different measurement setups for chemoresistors and electrochemical transistors is given in chapter 1.4.

1.1 Conducting polymers

The conductivity of π-conjugated polymers was discovered in 1977 by Heeger, MacDiarmid and Shirakawa.[6],[7] Since that time a huge number of publications reported about their synthesis, characterization and application in various fields. Typical monomers of conducting polymers are shown in Fig. 1.

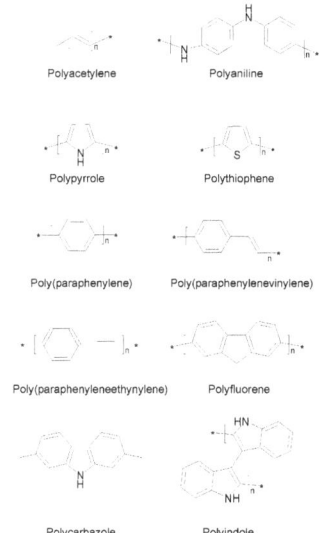

Figure 1. Main classes of conducting polymers

Introduction

The most fascinating property of conducting polymers is their intrinsic conductivity and the ability to switch this conductivity over 10 orders of magnitude.[6],[7] Conducting polymers show almost no conductivity in the neutral (uncharged) state. Their intrinsic conductivity results from the formation of charge carriers upon oxidizing (p-doping) or reducing (n-doping) their conjugated backbone. The more common oxidation can be explained with the band structure evolution shown in Fig. 2. According to Bredas et al. upon oxidation of the neutral polymer (a), relaxation processes causes the generation of localized electronic states and a polaron is formed (b).[8] If now an additional electron is removed, it is energetically more favourable to remove the second electron from the polaron than from another part of the polymer chain. This leads to the formation of one bipolaron rather than two polarons (c).[8] However it is important to note that before bipolaron formation the entire conducting polymer chain would first become saturated with polarons.[9] This model mainly based on spectroscopic data is widely accepted, however recently a model similar to redox-polymers was suggested on the basis of in-situ conductivity measurements.[10]

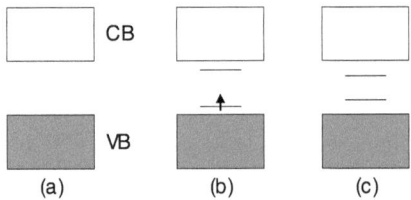

Figure 2. Band structure of conducting polymers in neutral state (a), after oxidation to polaron (b) and bipolaron (c) state.

The charge generation in the conducting polymer accompanied by the reversible intercalation of ions in the polymer matrix leads to significant changes in the optical, ionic, electrical and morphological properties of conducting polymers.[11] These properties changes can be tuned by using different dopants varying from small molecules to high molecular weight polymers as well as by using different preparation techniques. Table 1 shows several properties of conducting polymers that change upon altering their doping-state.

Introduction

Table 1. Qualitativ properties of CPs according to their charging state

Property	Reduced	Oxidized (P-doped)
Stoichiometry	Without anions (or with cations)	With anions (or without cations)
Content of solvent	Small	Higher
Volume	Increase with oxidation	
Colour: cathodically coloring	Transparent or bright	Dark
anodically coloring	Dark	Transparent
IR optical properties	Highly transmissive	Highly absorptive
Electronic conductivity	Semiconducting	Metallic
Ionic conductivity	Smaller	High
Diffusion of molecules	Dependent on structure	
Surface tension	Hydrophobic	Hydrophilic

Doping of conducting polymers can be done either chemically or electrochemically. In chemical doping the oxidation is accomplished by exposing the conducting polymer to oxidizing compounds like iodine. Another unique chemical doping procedure is the doping of Polyaniline (PANI) due to protonation.[12] This leads to an internal redox reaction converting the non-conducting form of PANI (emeraldine base) to a conductor (emeraldine salt).

Although chemical doping is an efficient process, controlling the level of dopant ions is rather difficult. Attempts to reach intermediate doping levels resulted in inhomogeneous doping. As an alternative, electrochemical doping allows a fine tuning of the doping level by simply adjusting the potential between the working and counter electrodes.[13] The working electrode supplies the redox charge to the conducting polymer, while ions diffuse in or out of the electroactive film to compensate the electronic charge. Thus any doping level can be achieved by setting the electrochemical cell to a desired potential and waiting for the system to attain an equilibrium state. This type of doping is permanent, meaning that the charge carriers remain in the film unless a neutralization potential is applied.

Conducting polymers can be synthesized either by addition of an external agent to the monomer solution (this approach is often referred as "chemical synthesis" of conducting polymers) or by electrochemical reaction.[14] Chemical synthesis of conducting polymers is usually performed by such oxidants as $NH_4S_2O_7$ or $FeCl_3$. Another approach is to couple functionalized monomers by coupling reactions like Stille or Suzuki coupling reaction.[15] Electrochemical synthesis is used for a direct deposition of non-soluble conducting polymer films on conducting substrates. An advantage of this method is the possibility to

Introduction

control the film thickness by the charge passed through the electrochemical cell during the film growth. Other popular techniques for depositing thin films on various substrates are drop, spin and spray coating from a solution of a chemically synthesized conducting polymer, the deposition of one or more monomolecular layers of conducting polymer by Langmuir-Blodgett-technique, or coating of substrates by bilayers of a conducting polymer and a opposed charged polymer by the Layer-by-Layer technique. Soluble conducting polymers can be synthesized by using side chain functionalized monomers for the synthesis. To obtain solutions of usually insoluble polymers, synthesis can be also carried out in the presence of surfactants or soluble polymers, which complex the insoluble polymer and keep it in solution. The most prominent example of this class of materials is the complex between polyethylenedioxythiophene and polystyrenesulfonate (PEDOT / PSS).[16]

1.1.1 Conducting polymer based sensors

There are several reasons to apply conducting polymers in chemo- and biosensors. All of them posses an intrinsic affinity towards redox active species and many to acidic or basic gases and solvent vapours.[1] Furthermore they can be modified with receptors to obtain a specific interaction with the analyte. Receptors can be covalently attached to the conjugated backbone of the conducting polymer, or physically entrapped in the polymer matrix. The class of receptors range from boronic acids, over crown ethers to DNA, proteins or metal or metaloxide nanoparticles.[1] Being immobilized on conducting polymers, such receptors provide an important advantage in comparison to monomer based receptors: the conducting polymer-wire provides a collective system response leading to high signal amplification in comparison to single molecular receptors.[17] Zhou et al. demonstrated that a conjugated polymeric receptor for methylviologen shows a 65 times signal amplification in comparison to the monomer based receptor.[18],[19] The amplification depends on the molecular weight.

There are several recent reviews on application of conducting polymers in sensors. A broad discussion of applications as well as the design of such sensors and the use of combinatorial techniques for evaluation of sensor materials is provided in [1]. A detailed review on chemical sensors based on amplifying fluorescent conjugated polymers was published by Thomas et al.[20] The detection of various analytes ranging from ions to proteins is discussed in this work. Conducting polymer based gas sensors are discussed in [21]. The application of conducting polymers in chemo- and biosensors can be realized with a number of different transducing techniques, allowing one to choose the most appropriate one for a particular sensor design. A detailed discussion of these aspects is given in [1].

1.2 Graphene

A flat monolayer of carbon atoms connected by sp^2-bonds into a two-dimensional (2D) honeycomb lattice, which is the building block for graphitic materials, is called graphene. Graphene has been studied theoretically, for sixty years, however only in 2004 the first free standing graphene layer was found,[22] and follow up experiments confirmed that its charge carriers are indeed mass less Dirac fermions,[23] as proposed earlier in theoretical studies. Since this time grapheme has attracted an enormous interest first in physics but in recent years also more and more in chemistry. There are several methods to produce graphene layers. The most simple and commonly used technique in physics is the micromechanical cleavage of graphite by repeated peeling of graphite with an adhesive tape[22] or by drawing with a piece of graphite.[24] However these methods produce only very few single layer graphene flakes hidden within a large quantity of thin graphite flakes, which have to be searched by optical microscopy. This is possible as graphene becomes visible in an optical microscope if placed on top of a Si wafer with a 300nm thick SiO_2 layer, by a feeble contrast with respect to an empty wafer.[24]

In another approach, films of single and few layer graphene can be grown on metal surfaces by chemical vapour deposition from hydrocarbon gases, such as methane, at temperatures of ca. 1000°C. [25],[26] It was demonstrates that thin nickel films and optimized cooling conditions can yield monolayer films. The properties of the graphene obtained by this technique can approach those of mechanically exfoliated graphene from highly ordered pyrolytic graphite (HOPG).[27] However the graphene formed on the metal substrates has to be transferred to an insulating surface before it can be used in electronic applications.

Another promising approach is the growth of epitaxial single or multilayer graphene by thermal sublimation of silizium from the surface of single crystalline SiC wafer at 1200 – 1500°C. [26],[28] In this process the removal of Si leaves carbon atoms on the surface, which reconstruct into graphene layers and grow continuously on the flat surface. The thickness of the graphene layer depends on the annealing time and the temperature.

Introduction

All the techniques described only produce graphene on surfaces. However for many applications it would be beneficial to have graphene in solution. This can be achieved by chemical exfoliation of graphite. Besides less used approaches to exfoliate graphene by ultrasound in special solvents[29] or by pretreating with oleum and tetrabultylammonium hydroxide,[30] the most used technique is the exfoliation by oxidizing graphite to graphiteoxide according to Hummers[31] or Staudenmeiers[32] method. By modifiying this method one can obtain stable dispersion of exfoliated single layer grapheneoxide. Graphiteoxide itself is studied since 1859 when Brodie oxidized graphite by a mixture of potassium chloride and fuming nitric acid.[33] This procedure was further impoved by Staudenmeier.[32] Around 60 years later, Hummers and Offemann developed a method involving concentrated sulphuric acid, sodium nitrate and potassium permanganate for the oxidation of graphite.[31] In this case diamanganese heptoxide is the active species in the oxidation. Graphiteoxide was further studied in the 1960 by Boehm. It was observed that by heating graphiteoxide, CO and CO_2 evolves already at room temperature leading to a darkening of graphiteoxide with time, but much faster at temperatures higher than 160°C. [34]-[36] Until this time several models have been developed to describe the structure of grapheneoxide. The most accepted now is the model by Lerf-Klinowski which assigns hydroxylic and expoxy groups as the main functional groups on grapheneoxide.[37] Furthermore carboxylic groups exist at the edges of the flakes. Grapheneoxide can be converted to graphene by reduction with hydrazine hydrate,[38] $NaBH_4$,[39] or ascorbate.[40] By carefully choosing the right conditions (pH > 8, low ionic strenght) during reduction stable dispersions of graphene in water can be obtained.[38]

The reduction removes most of the epoxy, hydroxyl and carboxylic groups on the grapheneoxide, yielding a conductive material. However graphene flakes prepared by this technique still have a lot defects and inferior properties than graphene prepared by mechanical exfoliation. A convenient method to improve the quality of the flakes is to heat them in an argon or argon / hydrogen atmosphere or under vacuum up to 1100°C. [41],[42] Furthermore chemical vapour deposition of ethylene at 800°C increases the conductivity of layers made from reduced grapheneoxide.[43]

1.2.1 Graphene in sensor application

Since its first detection graphene was suggested as ultrasensitive sensor material, allowing even single molecule detection.[44] A number of publications have investigated the use of graphene as a gas sensor material.[5] It was found that its conductivity is sensitivity towards NO_2,[44]-[46] NH_3,[40],[44],[45] water,[47],[48] 2,4-dinitrotoluene[45],[49] and solvent vapours.[47],[49] However it was also reported that the high sensitivity of graphene results probably from defects and / or impurities on graphene.[4] By treating graphene with an Ar / H_2 mixture at 400°C its sensitivity towards various analytes decreased.[47] It has to be noted that there is probably also a huge difference between different forms of graphene (e.g. chemically obtained vs. mechanically exfoliated or CVD deposited).[4] Furthermore it was shown that by modification of graphene its sensitivity and selectivity can be tuned.[5] For example decorating graphene with palladium or platinum nanoparticles yields a material sensitive to hydrogen.[50],[51] In addition to conductivity changes upon exposure towards gases it was shown that ionic strength,[52] pH[53],[54] and protein adsorption[54] can change the conductivity of graphene.

Besides chemoresistors and transistors based graphene sensors, reduced grapheneoxide was also used in voltammetric and amperometric sensors because of its electrocatalytical properties towards hydrogen peroxide, O_2, NADH, dopamine and other biological interesting compounds.[4],[55],[56] It was also shown that a direct electron transfer between glucose oxidase and reduced graphene oxide is possible.[56]

The fluorescence quenching of grapheneoxide was used in a DNA sensor.[57] A fluorophor containing single DNA strand binds on the graphenoxide. In the absence of the complementary strand the fluorophor is in proximity of the grapheneoxide and its fluorescence is quenched. However if a complementary strand binds to the single strand, the duplex is removed from the grapheneoxide and the fluorescence cannot be quenched anymore.

1.3   Metallic nanoparticles

Metal nanoparticles have unique properties relative to bulk metals.[2] For example, gold nanoparticles have a different colour and are much better catalysts as the bulk material. A number of approaches have been demonstrated for the synthesis of metal nanoparticles. Reduction of metal salts dissolved in appropriate solvents produces small metal particles of varying size distributions.[58] A variety of reducing agents have been employed for the reduction. These include alcohols, glycols, metal borohydrides, and certain specialized reagents. Such synthesis results in nanoparticles embedded in a layer of ligands or stabilizing agents, which prevent the aggregation of particles. The stabilizing agents

Introduction

employed include surfactants such as long-chain thiols or amines or monomeric or polymeric ligands such as citrate or polyvinylpyrrolidone (PVP).[58],[59] Successful nanoparticle synthesis has also been carried out employing soft templates such as reverse micelles.[60],[61]

1.3.1 Metallic nanoparticles in sensors

In electrochemical (bio-)sensors metallic nanoparticles are used to catalyze electrochemical reactions, to favour the electron transfer between biomolecules and the electrode and for the immobilisation of receptors.[62] Most works in this field was done using amperometric sensors, however composite materials of metallic nanoparticles and conducting polymers or carbon nanomaterials were also used in chemoresistors. In chemoresistors the metal nanoparticle usually works as analyte adsorption site. Adsorption and consecutive reaction of the analyte on the nanoparticle results in an electron transfer between the nanoparticle and the transducer layer which alters the conductivity of the later, by changing its amount of charge carriers. Using this principle it is possible to tune the sensitivity and selectivity of conducting polymers and carbon nanomaterial based chemoresistors.

The most common metallic nanoparticles employed in electrochemical sensors are gold, platinum, and palladium nanoparticles, however also silver and copper nanoparticles were used. There are several reviews dealing with the functionalisation of conducting polymers[2],[63] and carbon nanotubes[64] with metallic nanoparticles. The use of nanoparticles in electrochemical sensors is discussed in several reviews.[62],[65]-[68] An overview of conducting polymer nanocomposite materials in sensor applications including composites with metallic nanoparticles is provided in [2].

1.4 Conductometric sensors

Conductometric transducing of the sensor response is probably the most common method in chemo- and biosensors based on conducting polymers and carbon nanomaterials. There are several advantages of this transducing technique: (i) Small perturbations anywhere along the conjugated system can alter the conductance of the whole polymer chain, carbon nanotube or graphene flake. Therefore, this approach provides a higher sensitivity than other techniques based on modification of integral volume properties of the polymer, as for example electrochemical or colorimetric techniques.[69] (ii) Conductometric sensors can be realized with a simple setup which nevertheless allows high precision measurements. (iii) Conductometric chemosensitive measurements can be realized even with nanowires,[70] therefore this technique is perfectly compatible with the actual trend for miniaturization of analytical devices. (iv) Single chemoresistors can be easy combined into sensor arrays.   (iv) Using RFID technology, such sensors can be also adapted for non-contact measurements.[71]

There are several devices for measuring the conductometric response of such sensors. The simplest and most often used is a chemoresistor (Fig. 3 a, d, e). In the more common two point technique (Fig. 3 a) the conducting polymer is deposited between two (typically - interdigitated) electrodes separated by a narrow gap. The conductivity is measured by applying a constant current or voltage (dc or ac) between these electrodes and measuring the resulting voltage or current. The less used four-point measurement technique measures the conductance of the bulk polymer layer without an influence of the potential drop on the polymer – metal contacts (Fig. 3 d). This technique was modified recently by combining the two- and four-point techniques for simultaneous measurements (Fig. 3 e).[72],[73] Another possibility is the use of organic field effect transistors as sensors.[74]-[76] Here the current between the source and drain electrodes is controlled by the gate voltage. However these devices will not be discussed in here. Based on the similarity of the measurement configuration, Wrighton et al.[77]-[80] introduced the term "electrochemical transistor" for a device based on CP in which the redox state of this polymer is controlled by applying a voltage between the working electrodes (source and drain) and a reference electrode (gate) (Fig. 3 c, f).

The measurements of electrical current between two or more electrodes placed on the same solid support is certainly the most practicable configuration, however the low conductivity of some CP at typical conditions of bioanalytical applications (neutral pH, modest oxidation potential) may complicate this approach. In such cases, measurements of the resistance between an electrode coated by a conducting polymer and an electrode in the solution are used (Fig. 3 b). These measurements can be realized by two- or three-electrode circuits which are common in electrochemistry. An advantage of such measurements is the about 100 times higher ratio of the electrode area to the layer thickness (or an effective distance between electrodes placed on solid support), resulting in an about 100 times lower absolute resistance than for the measurements of lateral

Introduction

resistance.[1] All the configurations shown in Fig. 3 can be used in dc or ac mode and for impedance spectroscopy.

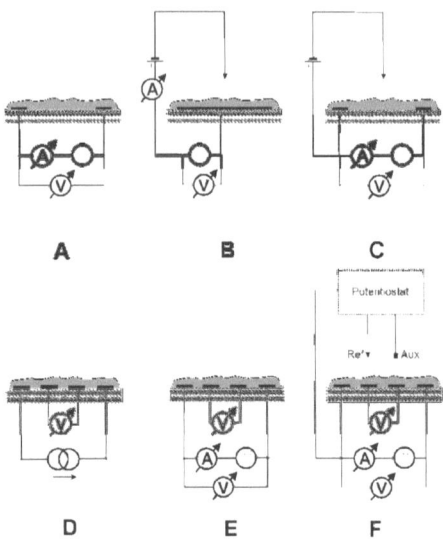

Figure 3. Main configurations used for analysis of resistance of conducting polymers. A: two point configuration without fixation of polymer potential. B: typical configuration used in electrochemical experiments. C: two point configuration with fixation of polymer potential. D: "classical" four-point technique with a current source. E: s24-configuration providing simultaneous two- and four-point measurements without fixation of polymer potential. F: s24-configuration with fixation of the electrode potential.

1.4.1 Chemoresistors

In most chemoresistors conductance changes are measured by the two point technique. A small dc probe voltage (~ 5-100 mV) or a constant current is applied between two electrodes separated by a small gap and the resulting current / voltage drop is measured. Such microelectrodes are mainly fabricated using photolithography. About 200 nm thick gold or platinum films are sputtered on oxidized silicon or glass wafers using a very thin adhesion layer. The gap between the electrodes is usually between 1.5 μm and 100 μm, however narrower gaps are favoured, as a smaller amount of polymer is needed to cover the gaps, resulting in an increased sensitivity. If the polymer is grown electrochemically over the gaps, larger gaps usually lead to thicker polymer films and therefore in a loss of sensitivity.[81],[82] Pre-treatment of the electrode support with e.g.

hydrophobic silanes can improve the lateral growth of the polymer over the gaps,[83]-[86] whereas pre-treatment of gold electrodes with thiol modified monomers can improve the contact between the electrode and the polymer film and therefore lower the influence of the contact resistance.[72] To achieve a longer gap length at limited electrode area, interdigitated electrodes are commonly used.

An application of a constant voltage or current to measure conductance changes in sensors based on conducting polymer leads to some problems. It may induce irreversible or slow-reversible changes in the polymer, which can be avoided by using of ac technique or by alteration of polarity of dc pulses.[87] In addition, passing a high current through the conducting polymer between the microelectrodes leads to a self-heating of the polymer, which makes its conductance sensitive to such parameters as e.g. air-flow.[88] The thermal response time was found to be in the order of 1 ms. Therefore the probe power should not exceed a certain limit. Another reason to keep the voltage probe as small as possible is the non-linearity of the current voltage dependence. In contrast, the signal to noise (S/N) ratio increases with increasing current. As a consequence, a compromise between a sufficiently high current to provide a suitable S/N ratio and the need to keep self-heating effects negligible has to be found.[88] Alternating current measurements have some advantages over direct current measurements. It was reported that the current noise during conductivity measurements displays flicker noise behaviour and decreases with 1/f (where f is frequency) with frequency increase. This noise behaviour was explained by a contribution of the contact resistance between single polymer grains to the overall resistance of the polymer film. This intergrain resistance is shunted by a capacitance bypassing the resistance at higher frequencies.[88] Furthermore, using ac measurements, one can make complete impedance analysis or monitor simultaneously resistive and capacitive changes which enhance the sensor selectivity.[89],[90] A sensor device interrogating the response at different frequencies was reported. Measurements at few selected frequencies are much faster than measuring the whole frequency spectrum by impedance spectroscopy.[91] Using multiple electrodes and multiplexing between the electrodes allows increasing the measurement throughput.

The conductance between closely spaced electrodes is provided mainly by a thin polymer layer close to the electrode substrate. Therefore thinner films are more sensitive towards vapours than thicker ones.[82],[92] In non-equilibrium state the analyte concentration decreases through the film thickness, therefore the polymer near the electrode support is less affected by analyte interaction in thicker films. This effect is expected to be especially

Introduction

strong if an interaction of polymer with analyte leads to an increase of the polymer resistance.

The resistance measured by the two-point technique includes the bulk polymer resistance and the resistance between the contacts and the polymer. If the contact resistance is high and shows in comparison to the polymer resistance only small changes upon analyte interaction, it can limit the sensitivity of the system. Most synthesis techniques leads to the formation of micrometer or submicrometer thick layers of conducting polymers, therefore the method of Cox and Strack[93] based on the variation of the ratio of the contact area to the material thickness can hardly be applied to distinguish bulk and contact resistances of conducting polymers. Principally, it can be done by impedance spectroscopy: by measuring a wide frequency spectra one can separate the resistance of the polymer and the resistance between the polymer and the metal contacts.[94],[95] However these measurements are relatively slow as a wide frequency range has to be covered. Additionally, the results of impedance spectroscopy are influenced by the selection of equivalent circuit used for the data analysis. More easily this problem can be solved by measurement techniques based on four-electrode configuration.

In the four-electrode configuration whose invention relates to Sir W. Thompson (Lord Kelvin), the conductivity is usually measured by applying a constant current between the two outer electrodes and by measuring the potential difference between the inner electrodes. This potential difference is measured by a high-impedance voltmeter, and therefore is not influenced by ohmic potential drop on the contact resistance of the inner electrodes.

If the contact resistance is so high that it limits the dynamic range of the sensor signal, the application of a four-electrode configuration results in a higher sensitivity.[81],[72] On the other hand four-electrode measurements alone provide no information on the resistance of polymer/metal contacts. However, many processes lead to a detaching of the polymer or to a formation of low conducting contacts due to modifications of the polymer band structure. This results in a higher contact resistance. On the other hand, the contact resistance can contain additional analytical information, which maybe useful in analyte discrimination.[96] The effects of contact resistance on the sensor sensitivity were investigated by Partrige et al. by multiplexing between four electrodes. They observed up to 55 % higher sensitivity for four-electrode configuration.[81]

Simultaneous two and four-point resistance measurements (s24) (Fig. 1 E), introduced by Kulikov et al., provide a possibility to measure both: four- and two-point resistance.[72],[73],[97] The ratio of the resistances measured by two- (R_2) and four-point (R_4) techniques provides a valuable information on the quality of metal/polymer contacts.

Several approaches to get a quantitative evaluation of the contact resistance were described. Similar to the s24 technique other techniques to measure the contact resistance are discussed in literature. It can be obtained by: (i) switching between a four- and a three electrode configurations.[98] (ii) Potential measurements between source and sensing electrodes and linear extrapolation of the voltage drop allows calculating the voltage drop at the source and drain contacts.[99] (iii) An investigation of the resistance dependence on the distance between two electrodes and determination of the contact resistance by subsequent extrapolation of this dependence to zero distance.[100],[101]

1.4.2 Electrochemical transistors

A number of applications require the control of the polymer redox state. This can be done electrically, by fixation of the polymer potential relative a reference electrode. Such measurements are often referred as in-situ resistance measurements and the measurement setup is called electrochemical transistor. This designation was accepted in literature and is therefore used here; however it should not be confused with semiconductor or organic transistors operating in a different way. The first application of electrochemical transistors for conducting polymers was described in 1984 by White et al. [77]

They used a symmetric configuration consisting of three gold electrodes 3 μm wide separated by a 1.4 μm gap placed on an insulation support. The middle electrode used to control the redox-state of the polymer was connected as a working electrode with a potentiostat. External auxiliary and reference electrodes were placed in the electrolyte solution. Due to the analogy to field effect transistors, the middle gold microelectrode was named gate and the two outer electrodes source and drain.

This configuration was further simplified by excluding the middle electrode and by control of the redox state of the polymer through the source (or drain) electrode.[78]-[80] This resulted in a narrower gap between the source and drain electrode and therefore a faster response time. However, at high polymer resistivity this asymmetric configuration may lead to deviation of the polymer redox state near the second electrode and therefore to inhomogeneous polymer properties between the source and drain. A further simplification of the configuration of electrochemical transistor can be performed the by the replacement

Introduction

of the potentiostat with reference and auxiliary electrodes by just a reference electrode connected through a potential source to the source (or drain) electrode. However in this case the ohmic drop at high gate currents can lead to deviations from the applied potential. A potential difference between source and drain electrodes required for conductivity measurements can be achieved by using a bipotentiostat.[102] Using a bipotentiostat, the potentials of the source and drain electrode (W1 and W2) are controlled simultaneously by keeping a constant small potential difference (e.g. 5 mV) between the two electrodes. The currents through drain and source electrodes consist of faradaic components caused by the redox reactions in the polymer and an ohmic component caused by the potential difference between the drain and source electrodes. Assuming that the magnitudes of the faradaic currents (i_F) at both electrodes are equal, one can calculate the ohmic current through the polymer film as $I_\Omega=(I_{W1}-I_{W2})/2$.[102] However the assumption that the faradaic currents are the same at both working electrodes is only true if the surface of both electrodes is exactly the same.

Another approach is based on combination of usual two- or four- electrode configurations for resistance measurement with a reference electrode connected directly or through a potentiostat. However if a constant voltage is applied between the source and drain electrodes, the measured current contains not only the current between the two electrodes but also the current between polymer layer and auxiliary or reference electrode in the electrolyte.[103] To overcome this problem, low frequency (< 1 Hz) voltage pulses[97],[103] or triangle voltage waveforms of alternating polarity are applied.[104] In these cases different methods to calculate the polymer resistance were used. Wrighton and coworkers calculate it from the slope of the I-V characteristic at zero voltage.[104] If pulses instead of triangle waveforms are used, the current is usually measured at the end of each pulse, when transient effects are minimal. In [103] the current is calculated from three successive measurements, by subtracting the negative pulse current from the average current of the two surrounding positive pulses. Averaging of the positive and negative pulse yields almost the same results. The elimination of the current crossing the electrochemical cell is particularly important if the current between source and drain is very small, e.g. if the polymer has a high resistance.

To use such electrochemical transistors in sensor applications, a detailed characterisation of factors influencing their in-situ resistance should be done first. Such in-situ resistance measurements have been widely used to study thin conducting polymer films.[78],[79],[86],[104]-[114] This approach was also combined with other techniques, like cyclic

Introduction

voltammetry,[108],[110],[111],[113],[115]-[117] electron spin resonance [110],[103] or quartz crystal microbalance.[115],[116]

The configuration of an electrochemical transistor has been used to fabricate pH-sensors,[84] ion-sensors,[118],[119] or for detection of redox active compounds.[79],[120] A modification of conducting polymers with enzymes which interact directly with the polymer or release/uptake during their enzymatic cycle any compounds effecting the polymer resistance is an approach to design conductometric enzymatic biosensors.[85],[121]-[123]

In a simple realization, external reference or reference / auxiliary electrodes can be used in such configuration; however these electrodes can also be implanted on the microchip surface. This is especially important in solid state devices. Solid state electrochemical transistors were first reported by Chao et al.[124],[125] They deposited a solid electrolyte over an array consisting of eight microelectrodes. Two of these electrodes were used as counter electrodes, the other six were connected by a conducting polymer film. A spot of silver glue was used as reference electrode. The chip was coated by polyvinylalcohol and suggested as a humidity sensor.

The influence of faradaic and non-faradaic processes at the gate electrode on the performance of electrochemical transistors was evaluated.[126] The charge needed for oxidation (doping) of a conducting polymer can be compensated by discharge of the double layer at the gate electrode (non-Faradaic process) or by a reduction process at the gate electrode (Faradaic process). However, the relatively small capacitance of the ionic double layer can limit the charge of the polymer oxidation. Faradaic processes have much higher pseudocapacitance, this leads to a higher sensitivity of the transistor current to towards changes of its gate potential. To enable a Faradaic process on the gate electrode, it can be also covered by a conducting polymer. A high surface ratio between gate and work electrodes decreases the switching time.

The switching time and the charge needed to switch the device also depend strongly on the amount of polymer necessary to bridge the gap between source and drain electrode. A switching time of 0.1 ms was reported for a polyaniline based device using a gap of 50 – 100 nm. This device was switched by an electric charge below 1 nC. For a similar device with 1.5 μm gap, about 100 times higher charge was required.[127] Devices with very small gaps were also used in sensors. By modifying of polyaniline with glucose oxidase, a biosensor for glucose designed for in-vivo applications was fabricated.[128]

Another approach to use electrochemical transistors as sensors is to use the gate electrode as receptor layer and the conducting polymer between the source and drain electrodes only as transducer. It has been shown that the reduction of hydrogen peroxide

Introduction

on a platinum gate electrode changes the conductance of a PEDOT layer deposited between source and drain electrode, if a constant potential is applied between gate and source electrode.[129],[130]

Solid state electrochemical transistors with conducting polymers were also fabricated on flexible substrates.[131],[132] In this case PEDOT / PSS is used as contact, channel and gate material. Fast switching rates at low humidity can be achieved using a very hygroscopic solid electrolyte consisting of polystyrenesulfonate (PSS), ethylenglycol, sorbitol and $LiClO_4$.[133] Nilsson et al. suggested several measurement configurations for such transistors. The simplest one is a three electrode electrochemical transistor which has a gate electrode referenced to the source electrode. A gel electrolyte covers the channel between the source and drain electrode and the gate electrode.[134],[135] A modification of this configuration can be achieved if the gate electrode is not referenced to the source but to a second gate electrode which is in contact with the channel.[131],[132] This is similar to the configuration of White et al. with a middle electrode as gate electrode between the source and drain electrodes. The four terminal electrochemical transistor was tested as a humidity sensor.[132] The conductance change of the channel upon applying a gate voltage of 1.2 V depends strongly on the humidity dependent conductance of the solid electrolyte (Nafion). The same principle was used for sensing of the ionic strength of an electrolyte, however a configuration using two three terminal electrochemical transistors, in which one served as reference, was used.[134]

1.5   Aim of the work

The aim of this work was the investigation of new measurement techniques and new (composite-)materials based on conducting polymers, chemically derived graphene and metallic nanoparticles for their use in chemiresistors and electrochemical transistors. The design of a chip containing not only four gold electrodes for resistance measurements, but additionally two more gold electrodes which can serve as counter and reference electrode, allows the integration of an electrochemical transistor on a small chip. This is proved in this work. Such electrochemical transistors can be used to speed up the regeneration of the conducting sensor layer. In an electrochemical transistor the conducting layer can however serve not only simultaneously as receptor and transducer, but can have only transducer function by combination with an analyte sensitive gate electrode, as demonstrated in chapter 3.4. New (composite-)materials allow one to increase sensitivity and as a result

Introduction

often also the selectivity of the conducting sensor film. Especially the electrocatalytic effects of metallic nanoparticles can be used to enhance the sensitivity towards specific analytes. New approaches to fabricate such composite materials are given in this work. Chemically derived graphene is a promising material for applications in (bio-)chemical sensors. The characterisation of this material and its application in gas sensors is described in chapter 3.7. By modification with metallic nanoparticles its sensitivity and selectivity can be tuned.

1.6   References

[1] U. Lange, N. Roznyatovskaya, V. Mirsky, Anal. Chim. Acta **2008**, 614, 1-26.
[2] D. W. Hatchett, M. Josowicz, Chem. Rev. **2008**, 108, 746-769.
[3] K. Balasubramanian, M. Burghard, Anal. Bioanal. Chem. **2006**, 385, 452-468.
[4] W. Yang, K. Ratinac, S. Ringer, P. Thordarson, J. Gooding, F. Braet, Angew. Chem. Intern. Ed. **2010**, 49, 2114-2138.
[5] K. R. Ratinac, W. Yang, S. P. Ringer, F. Braet, Env. Sci.Techn. **2010**, 44, 1167-1176.
[6] H. Shirakawa, E. Louis, A. MacDiarmid, C. Chiang, A. Heeger, J. Chem. Soc., Chem. Comm. **1977**, 1977, 578-580.
[7] C. Chiang, C. Fincher Jr, Y. Park, A. Heeger, H. Shirakawa, E. Louis, S. Gau, A. MacDiarmid, Phys. Rev. Lett. **1977**, 39, 1098-1101.
[8] J. L. Bredas, G. B. Street, Acc. Chem. Res. **1985**, 18, 309-315.
[9] P. Chandrasekhar, Conducting Polymers, Fundamentals and Applications: A Practical Approach, Springer Netherlands, **1999**.
[10] J. Heinze, B. A. Frontana-Uribe, S. Ludwigs, Chem. Rev. **2010**, DOI doi: 10.1021/cr900226k.
[11] G. Inzelt, M. Pineri, J. Schultze, M. Vorotyntsev, Electrochim. Acta **2000**, 45, 2403-2421.
[12] W. R. Salaneck, I. Lundström, W. Huang, A. G. Macdiarmid, Synth. Met. **1986**, 13, 291-297.
[13] A. Heeger, Angew. Chem. **2001**, 113, 2660-2682.
[14] G. Wallace, G. Spinks, L. Kane-Maguire, P. Teasdale, Conductive Electroactive Polymers: Intelligent Materials Systems, CRC, **2003**.
[15] M. Jeffries-El, R. McCullough in Handbook of Conducting Polymers 3th Edition, Ed.

Introduction

T. Skotheim, J. Reynolds, CRC Press, **2007**.
[16] S. Kirchmeyer, K. Reuter, J. C. Simpson in Handbook of Conducting Polymers 3th Edition, Eds. T. Skotheim, J. Reynolds, CRC Press, **2007**.
[17] T. M. Swager, Acc. Chem. Res. **1998**, 31, 201-207.
[18] Q. Zhou, T. M. Swager, J. Am. Chem. Soc. **1995**, 117, 7017-7018.
[19] Q. Zhou, T. M. Swager, J.Am.Chem.Soc. **1995**, 117, 12593-12602.
[20] S. W. Thomas, G. D. Joly, T. M. Swager, Chem. Rev. **2007**, 107, 1339-1386.
[21] H. Bai, G. Shi, Sensors **2007**, 7, 267-307.
[22] K. S. Novoselov, A. K. Geim, S. V. Morozov, D. Jiang, Y. Zhang, S. V. Dubonos, I. V. Grigorieva, A. A. Firsov, Science **2004**, 306, 666-669.
[23] K. S. Novoselov, A. K. Geim, S. V. Morozov, D. Jiang, M. I. Katsnelson, I. V. Grigorieva, S. V. Dubonos, A. A. Firsov, Nature **2005**, 438, 197-200.
[24] K. S. Novoselov, D. Jiang, F. Schedin, T. J. Booth, V. V. Khotkevich, S. V. Morozov, A. K. Geim, Proc.Natl. Acad Sci. U. S. A. **2005**, 102, 10451-10453.
[25] S. Unarunotai, Y. Murata, C. E. Chialvo, N. Mason, I. Petrov, R. G. Nuzzo, J. S. Moore, J. A. Rogers, Adv. Mater. **2010**, 22, 1072-1077.
[26] A. Reina, X. Jia, J. Ho, D. Nezich, H. Son, V. Bulovic, M. S. Dresselhaus, J. Kong, Nano Lett. **2009**, 9, 30-35.
[27] K. S. Kim, Y. Zhao, H. Jang, S. Y. Lee, J. M. Kim, K. S. Kim, J. Ahn, P. Kim, J. Choi, B. H. Hong, Nature **2009**, 457, 706-710.
[28] C. Berger, Z. Song, T. Li, X. Li, A. Y. Ogbazghi, R. Feng, Z. Dai, A. N. Marchenkov, E. H. Conrad, P. N. First, W. A. de Heer., J. Phys. Chem. B **2004**, 108, 19912-19916.
[29] Y. Hernandez, V. Nicolosi, M. Lotya, F. M. Blighe, Z. Sun, S. De, McGovernl. T., B. Holland, M. Byrne, Y. K. Gun'Ko, J. J. Boland, P. Niraj, G. Duesberg, S. Krishnamurthy, R. Goodhue, J. Hutchison, V. Scardaci, A. C. Ferrari, J. N. Coleman, Nat Nano **2008**, 3, 563-568.
[30] X. Li, G. Zhang, X. Bai, X. Sun, X. Wang, E. Wang, H. Dai, Nat Nano **2008**, 3, 538-542.
[31] W. S. Hummers, R. E. Offeman, J. Am. Chem. Soc. **1958**, 80, 1339.
[32] L. Staudenmaier, Ber. Dtsch. Chem. Ges. **1898**, 31, 1481-1487.
[33] B. Brodie, Philos. Trans. R. Soc. London **1859**, 149, 249-259.
[34] H. Boehm, W. Scholz, Z. Anorg. Allg. Chem. **1965**, 335, 74-79.
[35] W. Scholz, H. P. Boehm, Z. Anorg. Allg. Chem. **1964**, 331, 129-132.
[36] W. Scholz, H. -. Boehm, Naturwissenschaften **1964**, 51, 160.

[37] A. Lerf, H. He, M. Forster, J. Klinowski, J. Phys. Chem. B **1998**, 102, 4477-4482.
[38] D. Li, M. B. Muller, S. Gilje, R. B. Kaner, G. G. Wallace, Nat. Nano **2008**, 3, 101-105.
[39] H. Shin, K. K. Kim, A. Benayad, S. Yoon, H. K. Park, I. Jung, M. H. Jin, H. Jeong, J. M. Kim, J. Choi, u. a., Adv. Funct. Mater. **2009**, 19, 1987-1992.
[40] V. Dua, S. Surwade, S. Ammu, S. Agnihotra, S. Jain, K. Roberts, S. Park, R. Ruoff, S. Manohar, Angew. Chem. **2010**, 122, 2200-2203.
[41] W. Gao, L. Alemany, L. Ci, P. Ajayan, Nat. Chem. **2009**, 1, 403-408.
[42] H. Becerril, J. Mao, Z. Liu, R. Stoltenberg, Z. Bao, Y. Chen, ACS Nano **2008**, 2, 463-470.
[43] V. López, R. Sundaram, C. Gómez-Navarro, D. Olea, M. Burghard, J. Gómez-Herrero, F. Zamora, K. Kern, Adv. Mater. **2009**, 21, 4683-4686.
[44] F. Schedin, A. K. Geim, S. V. Morozov, E. W. Hill, P. Blake, M. I. Katsnelson, K. S. Novoselov, Nat. Mater. **2007**, 6, 652-655.
[45] J. D. Fowler, M. J. Allen, V. C. Tung, Y. Yang, R. B. Kaner, B. H. Weiller, ACS Nano **2009**, 3, 301-306.
[46] G. Lu, L. E. Ocola, J. Chen, Appl. Phys. Lett. **2009**, 94, 083111-3.
[47] Y. Dan, Y. Lu, N. J. Kybert, Z. Luo, A. T. C. Johnson, Nano Lett. **2009**, 9, 1472-1475.
[48] I. Jung, D. Dikin, S. Park, W. Cai, S. L. Mielke, R. S. Ruoff, J. Phys. Chem. C **2008**, 112, 20264-20268.
[49] J. T. Robinson, F. K. Perkins, E. S. Snow, Z. Wei, P. E. Sheehan, Nano Lett. **2008**, 8, 3137-3140.
[50] R. S. Sundaram, C. Gomez-Navarro, K. Balasubramanian, M. Burghard, K. Kern, Adv.Mater. **2008**, 20, 3050-3053.
[51] A. Kaniyoor, R. Jafri, T. Arockiadoss, S. Ramaprabhu, Nanoscale **2009**, 1, 382-386.
[52] P. K. Ang, S. Wang, Q. Bao, J. T. L. Thong, K. P. Loh, ACS Nano **2009**, 3, 3587-3594.
[53] P. K. Ang, W. Chen, A. T. S. Wee, K. P. Loh, J. Am. Chem. Soc. **2008**, 130, 14392-14393.
[54] Y. Ohno, K. Maehashi, Y. Yamashiro, K. Matsumoto, Nano Lett. **2009**, 9, 3318-3322.
[55] M. Zhou, Y. Zhai, S. Dong, Anal. Chem. **2009**, 81, 5603-5613.
[56] C. Shan, H. Yang, J. Song, D. Han, A. Ivaska, L. Niu, Anal. Chem. **2009**, 81, 2378-2382.
[57] S. He, B. Song, D. Li, C. Zhu, W. Qi, Y. Wen, L. Wang, S. Song, H. Fang, C. Fan, Adv. Funct. Mater. **2010**, 20, 453-459.
[58] G. Schmid, Nanoparticles: From Theory to Application, Wiley-VCH, **2004**.

Introduction

[59] C. Rao, G. Kulkarni, P. Thomas, in Metal-Polymer Nanocomposites, Eds. L. Nicolais, G. Carotenuto, Wiley, **2005**.
[60] M. P. Pileni, J. Phys. Chem. **1993**, 97, 6961-6973.
[61] M. P. Pileni, Langmuir **1997**, 13, 3266-3276.
[62] X. Luo, A. Morrin, A. Killard, M. Smyth, Electroanal. **2006**, 18, 319-326.
[63] A. Malinauskas, J. Malinauskiene, A. Ramanavi ius, Nanotechn. **2005**, 16, R51.
[64] G. Wildgoose, C. Banks, R. Compton, Small **2006**, 2, 182-193.
[65] C. Welch, R. Compton, Anal. Bioanal. Chem. **2006**, 384, 601-619.
[66] E. Katz, I. Willner, J. Wang, Electroanal. **2004**, 16, 19-44.
[67] D. Hernández-Santos, M. González-García, A. García, Electroanal. **2002**, 14, 1225-1235.
[68] G. Wildgoose, C. Banks, H. Leventis, R. Compton, Microchim. Acta **2006**, 152, 187-214.
[69] K. Sugiyasu, T. M. Swager, Bull. Chem. Soc. Jpn. **2007**, 80, 2074-2083.
[70] A. Wanekaya, W. Chen, N. Myung, A. Mulchandani, Electroanal. **2006**, 18, 533-550.
[71] R. A. Potyrailo, W. G. Morris, Anal. Chem. **2007**, 79, 45-51.
[72] Q. Hao, V. Kulikov, V. M. Mirsky, Sens. Actuators BI **2003**, 94, 352-357.
[73] V. Kulikov, V. M. Mirsky, T. L. Delaney, D. Donoval, A. W. Koch, O. S. Wolfbeis, Meas. Sci.Technol. **2005**, 16, 95-99.
[74] J. Mabeck, G. Malliaras, Anal. Bioanal. Chem. **2006**, 384, 343-353.
[75] J. Janata, M. Josowicz, Nat. Mater. **2003**, 2, 19-24.
[76] C. Bartic, G. Borghs, Anal. Bioanal. Chem. **2006**, 384, 354-365.
[77] H. S. White, G. P. Kittlesen, M. S. Wrighton, J. Am. Chem. Soc. **1984**, 106, 5375-5377.
[78] D. Ofer, R. Crooks, M. Wrighton, J.Am.Chem.Soc. **1990**, 112, 7869-7879.
[79] J. W. Thackeray, H. S. White, M. S. Wrighton, J. Phys. Chem. **1985**, 89, 5133-5140.
[80] M. Wrighton, J. Thackeray, M. Natan, D. Smith, G. Lane, D. Belanger, W. Albery, Philos.Trans. R. Soc. London B, Biol. Sci. **1987**, 316, 13-30.
[81] A. Partridge, P. Harris, M. Andrews, Analyst **1996**, 121, 1349-1353.
[82] E. Stussi, R. Stella, D. De Rossi, Sens. Actuators B **1997**, 43, 180-185.
[83] M. Nishizawa, Y. Miwa, T. Matsue, I. Uchida, J. Electroanal. Chem. **1994**, 371, 273-275.
[84] M. Nishizawa, T. Matsue, I. Uchida, Sens. Actuators B **1993**, 13, 53-56.
[85] M. Nishizawa, T. Matsue, I. Uchida, Anal. Chem. **1992**, 64, 2642-2644.
[86] T. Matsue, M. Nishizawa, I. Uchida, Nippon Kagaku Kaishi **1995**, 493-501.

[87] A. Guiseppi-Elie, G. Wallace, T. Matsue, in Handbook of Conducting Polymers, CRC, **1998**.

[88] P. Harris, W. Arnold, M. Andrews, A. Partridge, Sens. Actuators B **1997**, 42, 177-184.

[89] F. Musio, M. C. Ferrara, Sens. Actuators B **1997**, 41, 97-103.

[90] M. Hassan Amrani, P. A. Payne, K. C. Persaud, Sens. Actuators B **1996**, 33, 137-141.

[91] M. Amrani, R. Dowdeswell, P. Payne, K. Persaud, Sens. Actuators B **1998**, 47, 118-124.

[92] P. N. Bartlett, S. K. Ling-Chung, Sens. Actuators **1989**, 19, 141-150.

[93] R. Cox, H. Strack, Solid-State Electron. **1967**, 10, 1213-1214, IN7-IN8, 1215-1218.

[94] C. Deslouis, M. Musiani, B. Tribollet, M. Vorotyntsev, J. Electrochem. Soc. **1995**, 142, 1902-1908.

[95] X. Ren, P. G. Pickup, J. Electroanal. Chem. **1997**, 420, 251-257.

[96] M. Krondak, G. Broncovic, S. Anikin, A. Merz, V. Mirsky, J. Solid State Electrochem. **2006**, 10, 185-191.

[97] U. Lange, V. M. Mirsky, J. Electroanal. Chem. **2008**, 622, 246-251.

[98] S. G. Haupt, D. R. Riley, J. Zhao, J. T. McDevitt, J.Phys.Chem. **1993**, 97, 7796-7799.

[99] P. V. Pesavento, R. J. Chesterfield, C. R. Newman, C. Frisbie, J. Appl. Phys. **2004**, 96, 7312-7324.

[100] Q. J. Cai, M. B. Chan-Park, J. Zhang, Y. Gan, C. M. Li, T. P. Chen, B. S. Ong, Org. Electron. **2008**, 9, 14-20.

[101] P. V. Necliudov, M. S. Shur, D. J. Gundlach, T. N. Jackson, Solid-State Electron. **2003**, 47, 259-262.

[102] M. Nishizawa, I. Uchida, Electrochim. Acta **1999**, 44, 3629-3637.

[103] J. Kruszka, M. Nechtschein, C. Santier, Rev. Sci. Instrum. **1991**, 62, 695-699.

[104] E. W. Paul, A. J. Ricco, M. S. Wrighton, J. Phys.Chem. **1985**, 89, 1441-1447.

[105] M. Gholamian, T. Kumar, A. Contractor, Proc. Ind. Acad. Sci., Chem.Sci. **1986**, 97, 457-464.

[106] L. Groenendaal, G. Zotti, F. Jonas, Synth. Met. **2001**, 118, 105-109.

[107] R. Holze, J. Lippe, Synth.Met. **1990**, 38, 99-105.

[108] M. C. Morvant, J. R. Reynolds, Synth. Met. **1998**, 92, 57-61.

[109] G. Schiavon, S. Sitran, G. Zotti, Synth.Met. **1989**, 32, 209-217.

[110] G. Zotti, Synth. Met. **1998**, 97, 267-272.

Introduction

[111] G. Zotti, G. Schiavon, Synth. Met. **1990**, 39, 183-190.
[112] E. Csahok, E. Vieil, G. Inzelt, J. Electroanal. Chem. **2000**, 482, 168-177.
[113] L. Olmedo, I. Chanteloube, A. Germain, M. Petit, E. Genies, Synth. Met. **1989**, 30, 159-172.
[114] H. Huang, P. Pickup, Acta Polym. **1997**, 48, 455-457.
[115] E. Sezer, M. Skompska, J. Heinze, Electrochim. Acta **2008**, 53, 4958-4968.
[116] E. Sezer, J. Heinze, Electrochim. Acta **2006**, 51, 3668-3673.
[117] W. Zhang, S. Dong, Electrochim. Acta **1993**, 38, 441-445.
[118] Z. Mousavi, A. Ekholm, J. Bobacka, A. Ivaska, Electroanal. **2009**, 21, 472-479.
[119] D. A. Bernards, G. G. Malliaras, G. E. Toombes, S. M. Gruner, Appl. Phys. Lett. **2006**, 89, 053505-3.
[120] P. N. Bartlett, J. H. Wang, E. N. K. Wallace, Chem. Commun. **1996**.
[121] P. Bartlett, P. Birkin, J. Wang, F. Palmisano, G. De Benedetto, Anal. Chem. **1998**, 70, 3685-3694.
[122] P. Bartlett, Y. Astier, Chem. Comm. **2000**, 2000, 105-112.
[123] M. Nikolou, G. G. Malliaras, Chem. Rec. **2008**, 8, 13-22.
[124] S. Chao, M. Wrighton, J. Am. Chem. Soc. **1987**, 109, 2197-2199.
[125] S. Chao, M. S. Wrighton, J. Am. Chem. Soc. **1987**, 109, 6627-6631.
[126] F. Lin, M. C. Lonergan, Appl. Phys. Lett. **2006**, 88, 133507-3.
[127] E. T. Jones, O. M. Chyan, M. S. Wrighton, J. Am. Chem. Soc. **1987**, 109, 5526-5528.
[128] E. S. Forzani, H. Zhang, L. A. Nagahara, I. Amlani, R. Tsui, N. Tao, Nano Lett. **2004**, 4, 1785-1788.
[129] D. J. Macaya, M. Nikolou, S. Takamatsu, J. T. Mabeck, R. M. Owens, G. G. Malliaras, Sens. Actuators Bl **2007**, 123, 374-378.
[130] Z. Zhu, J. T. Mabeck, C. Zhu, N. C. Cady, C. A. Batt, G. G. Malliaras, Chem. Commun. Y1 – 2004, 1556-1557.
[131] D. Nilsson, M. Chen, T. Kugler, T. Remonen, M. Armgarth, M. Berggren, Adv. Mater. **2002**, 14, 51-54.
[132] D. Nilsson, T. Kugler, P. O. Svensson, M. Berggren, Sens. Actuators B **2002**, 86, 193-197.
[133] M. Hamedi, R. Forchheimer, O. Inganas, Nat. Mater. **2007**, 6, 357-362.
[134] P. Svensson, D. Nilsson, R. Forchheimer, M. Berggren, Appl. Phys. Lett. **2008**, 93.
[135] D. Nilsson, N. Robinson, M. Berggren, R. Forchheimer, Adv. Mater. **2005**, 17, 353-358.

2. Methods

2.1 In-situ simultaneous two- and four-point measurement

2.1.1 Theory and working principle

The simultaneous two- and four-point technique for conductivity measurements was suggested by V. M. Mirsky and V. Kulikov in 2003 and used for conductivity measurements of conducting polymers.[1],[2],[3]

The setup of the in-situ simultaneous two- and four-point conductivity measurement technique is shown in Fig. 4. An electrode consisting of four stripes, covered with a conducting film is connected to a Keithley source meter (K2400), which applies a small alternating voltage (V_D) between the outer two electrodes, and measures the resulting two point current. Typically this voltage is in the range between 10 mV to 50 mV. Furthermore the two inner electrodes are connected to a high-impedance Keithley voltmeter (K617), which measures the four-point voltage drop. To control the potential of the measurement electrodes one of the four bands and a reference electrode are connected to a voltage source (K617) or one band is connected as working electrode to a potentiostat. In this case potential control is achieved with the help of a reference and a counter electrode.

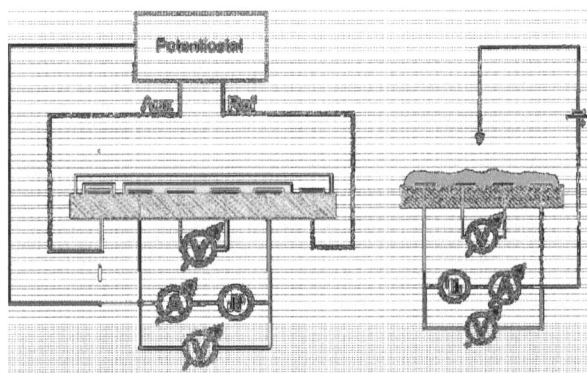

Figure 4. Set-up for the in-situ simultaneous two- and four-point measurement with reference and counter electrode on the chip (left) and with external reference electrode in electrolyte solution (right).

Methods

In Fig. 5 the scheme of the alternating pulse mode is presented. This mode was chosen to avoid polarization effects of the layer. The duration of one pulse is 350 ms.

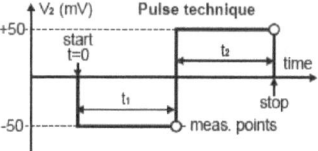

Figure 5. Scheme of the pulse mode used for simultaneous two- and four point measurement. The current is measured at the end of each pulse (indicated by a circle).

Alternating pulses are especially important if the setup of an electrochemcial transistor is used. In this case the measured current contains not only the current through the polymer film (I_D), but also the current between the film and the counter or reference electrode (I_F). Especially if the film resistance is high this can lead to significant measurement errors. The current measured during the positive voltage pulse can be therefore written as:

$$I^+ = I_D + I_F$$

And for the negative pulse as:

$$I^- = -I_D + I_F$$

By assuming that the current I_F is the same for the positive and the negative pulse, this current can be eliminated by averaging of the positive and negative pulse:

$$\frac{I^+ + (-I^-)}{2} = \frac{I_D + I_F + I_D - I_F}{2} = \frac{2I_D}{2}$$

Using the average current one can calculate the two point resistance according to:

$$R_2 = \frac{V_D}{I_D}$$

Furthermore the average four-point voltage drop is calculated from the voltage drop of the two pulses (V^+, V^-), measured with the K617 according to:

$$V_4 = \frac{V^+ + (-V^-)}{2}$$

Methods

V_4 is used for the calculation of the four point resistance:

$$R_4 = \frac{V_4}{I_D}$$

Additionally e the two- and four-point resistance of each pulse polarity are also monitored:

$$R_2^+ = \frac{U}{I^+}; \ R_2^- = \frac{-U}{I^-}; \ R_4^+ = \frac{U^+}{I^+}; \ R_4^- = \frac{U^-}{I^-}$$

All six calcultated resistances, the measurement time and the potential of the reference electrode are written into a data file. The computer program for the two- and four-point resistance measurements was developed by V.Kulikov, based on software of Agilent Vee.

The advantage of the simultaneous two- and four-point technique over conventional conductivity measurement techniques is the possibility to obtain information about the bulk resistance of the polymer layer as well as the contact resistance by a single measurement. The resistance R_2, measured by the two-point technique, is a sum of resistance of contacts R_c and bulk resistance of the polymer between the two outer electrodes R_L: $R_2 = R_c + R_L$. In contrary the resistance R_4 measured by the four-point technique is the bulk resistance of the polymer between the two inner electrodes (Fig. 8). From geometric consideration, for thin (less than gap) polymer layers, the part of the polymer resistance between the inner electrodes ($\alpha = R_4/R_L$) should be between 1/9 and 3/5. Therefore, if the ratio $R_2/R_4 > 9$, we can conclude a measurable contribution of the contact resistance into R_2. The R_2 to R_4 ratio therefore can work as an internal quality control of the sensor. If there is an increase in R_2 accompanied with an increase in the ratio R_2 to R_4, the resistance increase is not due to an analyte interaction with the chemosensitive film, but due to changes in the quality of the contacts.

2.1.2 Electrodes

150 nm thin film gold electrodes prepared by photolithographic lift-off technique on a glass wafer with a thickness of 0.5 mm were provided by Fraunhofer IZM-München. A 15 nm TiW sub-layer was used to enhance the adhesion of gold on the

Methods

glass. The electrodes consist of six bands. The two outer broad bands serve as counter and reference electrodes, whereas the inner four bands are used for simultaneous two- and four-point measurements. For the inner electrodes either linear or interdigitated bands are used (Fig. 6). The working area of the four inner electrodes was 0.574 mm² for the linear and 0.64 mm² for the interdigitated structure.

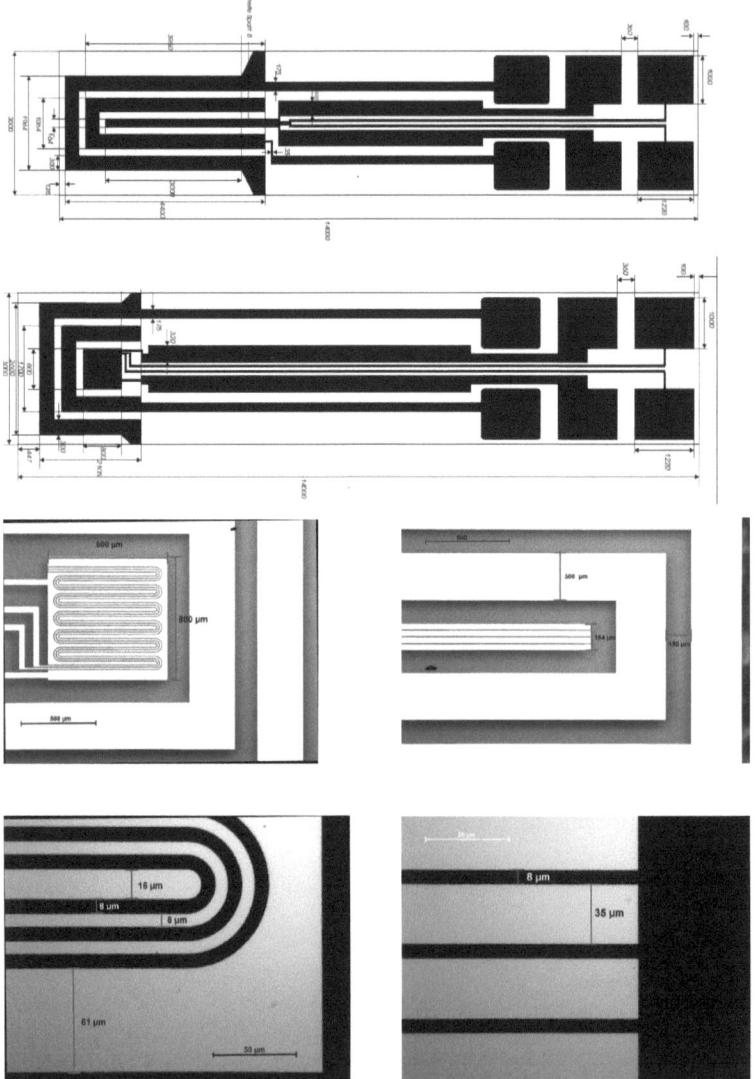

Figure 6. Electrode design and pictures of the sensor spot for the interdigitated (left) and linear electrodes (right)

It has to be noted that in some applications described in this work similar electrodes with two-or four-bands on SiO_2 separated by a 5 μm gap were used. A describtion of these electrodes is provided in [4].

2.2   References

[1] Q. Hao, V. Kulikov, V. M. Mirsky, Sens. Actuators B **2003**, 94, 352-357.
[2] V. Kulikov, V. M. Mirsky, T. L. Delaney, D. Donoval, A. W. Koch, O. S. Wolfbeis, Meas. Sci. Technol. **2005**, 16, 95-99.
[3] M. Krondak, G. Broncova, S. Anikin, A. Merz, V. Mirsky, J. Solid State Electrochem. **2006**, 10, 185-191.
[4]. V.Kulikov, Dissertation, Technische Universität München **2004**.

3. Results and Discussion

3.1. Simultaneous measurements of bulk and contact resistance

To evaluate an application of the simultaneous two- and four-point technique for a separated quantitative characterisation of the bulk polymer resistance and the polymer – metal contact resistance, polypyrrole (PPy) films on interdigitated and single spot gold electrodes were investigated by this technique and by impedance spectroscopy.

Analysis of the impedance spectra using an equivalent circuit is a possibility to obtain the contact resistance between the metal and the conducting polymer (see chapter 1.4.1).[1],[2] Therefore this method was chosen to evaluate the data obtained by simultaneous two and four-point measurements.

3.1.1. Results and discussion

The resistance measured by the 2-point technique gives the value R_2 which is the sum of the bulk film resistance and two contact resistances, while the resistance R_4 measured by the 4-point technique is the film resistance without contact resistances. Using the interdigitated electrodes shown in Fig. 6, R_4 is measured between some mean points of the inner electrode. The contact resistance can therefore be calculated as:

$$R_c = R_2 - \alpha \cdot R_4, \qquad (1)$$

where α is the geometrical factor. Taking into account the electrode geometry (Fig. 7), one can expect that the geometric factor α corresponding to the ratio of the bulk polymer resistance between two outer electrodes to the bulk polymer resistance between two inner electrodes, is between 5/3 and 9.[3]

The current from the outer electrodes flows in both directions, therefore the effective mean coordinates of the outer electrode are placed at 1/4 of its width from the inner part while the mean value of each inner electrode is in its middle (Fig. 7). This decreases the effective distance between the outer electrodes to 30 μm whereas the effective distance between the inner electrodes is 10 μm. Therefore, for thin homogenous films, whose thickness is much less than the width and gap of the electrodes, the factor α is about 3.

Several hundreds of measurements performed by combinatorial technique on electrode arrays covered by polyaniline as well as binary copolymers of polyaniline derivates with polyaniline showed that the ratio of R_2/R_4 measured at the conditions corresponding to the minimum value of this ratio, is 3.1±0.1.[4],[5] One can assume that such measurements conditions correspond to the case when $R_c \ll R_4$, therefore it was no contribution of the contact resistance into the measured R_2 resistance. The obtained value corresponds exactly to the value of α obtained from geometrical consideration for thin and homogeneous polymer films. Taking into account that the electrode geometry is the same for the all studied polymers, the polymer thickness is also very similar, and assuming that all polymer films are electrically homogeneous, one can use the equation (1) to calculate contact resistance from the values of R_2 and R_4 postulating α = 3. Below also the influence of a deviation of α from this postulate will be analyzed.

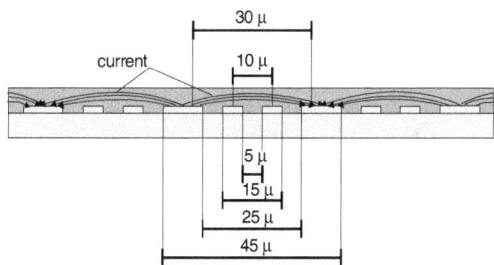

Figure 7. Simplified electrode geometry of a four contact electrode. The factor α corresponds to the ratio of effective distances between the outer and inner electrodes. For thin homogeneous polymer layers α = 30 μm /10 μm = 3.

Equation 1 allows to obtain direct information about the bulk polymer and the contact resistance from a single measurement. This was used for measuring the potential dependent resistance of PPy on interdigitated gold electrodes. The contact resistance between the PPy-film and the gold electrode was calculated from the potential dependent R_2 and R_4 values according to (1), by using the factors α = 5/3, α = 3 and α = 3.3 (Fig. 8). The value 5/3 corresponds to the theoretically lowest possible value of α, whereas the value 3.3 is close to the lowest value found for R_2 /

Results and Discussion

R_4 in this measurement. The curves for $\alpha = 5/3$ and for $\alpha = 3.3$ represent therefore the highest and lowest possible contact resistances at each potential. The results demonstrate that the consideration of maximal possible variations of α leads to variation of the calculated contact resistance for the factor of up to one order of magnitude but almost does not influence the behavior of the curve. For further data analysis the value $\alpha = 3$ corresponding to theoretical value for thin homogeneous polymer layers and experimentally confirmed by many hundreds of combinatorial measurements was used.

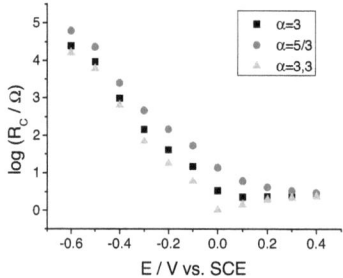

Figure 8. Potential dependent contact resistance of a PPy-film on interdigitated gold electrode at pH 2 calculated for three different values of α

Fig. 9 A shows the potential dependent contact resistance in comparison with the bulk polymer resistance measured by the four-point technique. From -0.6 V to 0 V both resistances show almost an identical behaviour. However from 0 V to 0.4 V the contact resistance remains constant while the bulk polymer resistance drops further. Literature data for the potential dependence of PPy resistance obtained by two-point measurement techniques[6],[7],[8],[9] are very close to our two-point data presented in the Fig. 9 B. A comparison with Fig. 9 A however demonstrates that although the behaviour of this curve describes qualitatively the bulk polymer properties at low potentials, it does not characterise the polymer behaviour at high potentials where the contact resistance prevails.

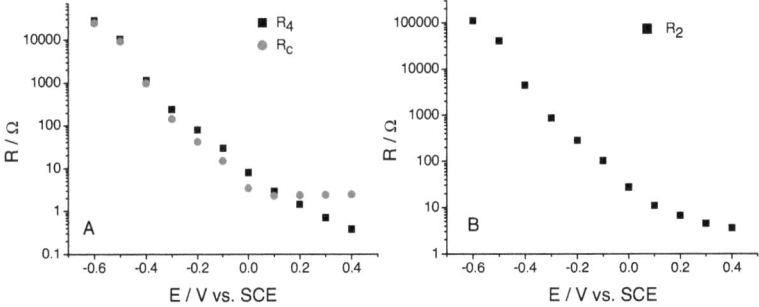

Figure 9. Potential dependence of two-point resistance (R_2), four point resistance (R_4) and contact resistance R_c of PPy-film at pH 2.

An alternative approach to analyse the contact resistance is based on impedance spectroscopy.[1],[2],[10] This approach was used for the validation of the measurements obtained by simultaneous two- and four-point measurement. The impedance spectra of PPy films were recorded on gold electrodes in the electrode potential range from -0.6 V to 0.4 V at 0.01 Hz – 60 kHz. The impedance spectra were described by the equivalent circuit shown in Fig. 10. The calculated curves presented in the Fig. 11 demonstrate that this equivalent circuit provides a good fitting of the experimental data. In the equivalent circuit the resistance R_c represents the contact resistance at the electrode / polymer interface, the capacitor C_c describes the capacitance of this interface. The resistance R_s is the uncompensated solution resistance and the serially connected two constants phase elements (CPE) each of them being shunted by a resistor (R_p and R_{ps} correspondingly) describe volume properties of the conducting polymer.[11] An exact description of the physical processes in the bulk polymer phase which are associated with these CPEs is not important for this aim, however one of these elements is probably associated with the bulk Faradaic pseudocapacitance[11] (it can be also approximated by a capacitor) while the second one is related to a diffusion processes (it can be also approximated by Warburg impedance). This interpretation is confirmed by the obtained values of the power parameter of these constant phase elements: depending on the redox-state of the polymer, this parameter for the first CPE is 0.82 - 0.95 which is close to its value for

Results and Discussion

an ideal capacitor (1.0) while for the second CPE this parameter is between 0.48 and 0.68 which is close to its value for a Warburg impedance (0.5).

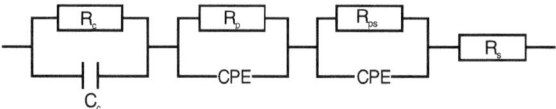

Figure 10. Equivalent circuit used for analysis of the impedance data.

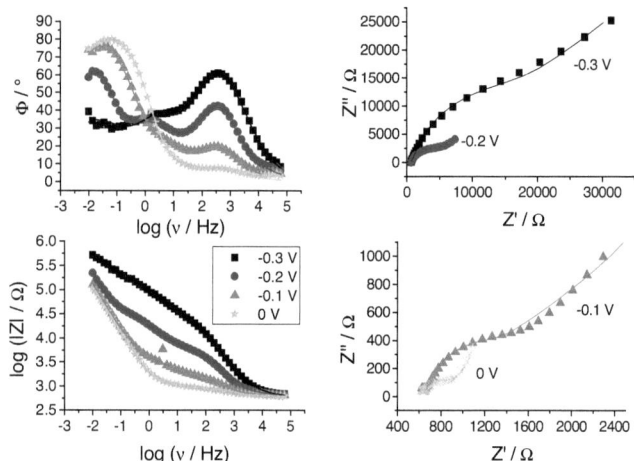

Figure 11 Bode- (left) and Nyquist (right) plots for a PPy film on a gold electrode at different potentials. Measurements were done from 0.01 Hz to 60 kHz, Bode plots are shown for the whole frequency range, Nyquist plots for the range 8–60 kHz. Data were fitted with equivalent circuit shown in Fig.10.

The high-frequency behaviour (i.e., the values of R_c and C_c) characterises electrical properties of an interface. According to [1],[2],[10] one can assume that this interface is the metal / polymer interface. Therefore the resistance obtained from fitting the high-frequency properties probably is the contact resistance between PPy and the gold electrode. Fig. 12 shows the plot of the normalized contact resistance obtained by impedance spectroscopy at different electrode potentials in comparison to the normalized contact resistance measured by simultaneous two- and four-point measurement technique. It can be seen that both values drop down between -0.6 V

Results and Discussion

and 0.1 V in about the same magnitude. Therefore, the value R_c calculated from the equation (1) seems to describe contact resistance at the polymer / metal interface.

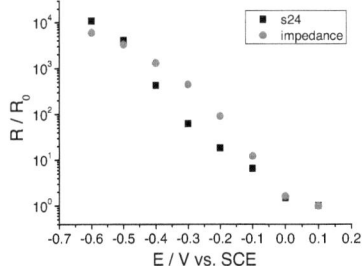

Figure 12. Comparison of the contact resistances determined by simultaneous two- and four-point technique (s24) and by impedance analysis. The contact resistance from s24 technique was calculated according to Eq. (1) by using $\alpha = 3$ (see text for details). The curves were normalized for the resistance at 0.1 V.

3.1.2. Experimental

Polypyrrole films were prepared from a 0.1 M solution of freshly distilled pyrrole in acetonitrile (Baker, HPLC grade) containing 0.1 M tetrabutylammonium hexafluorophosphate (NBu$_4$PF$_6$, Fluka, electrochemical grade) and 1% water, by cycling the potential between -0.2 V and 1.1 V. To obtain thin films (thickness ~ 1 μm) which cover the whole electrode area, five cycles were applied.

Impedance measurements were performed with disk gold electrodes deposited on the surface of silicon oxide coated silicon wafer with a working area of 0.38 mm^2.

The thickness of the films was estimated by a shift of the focus plane of an optical microscope. Electrochemical measurements were done using an Autolab PGSTAT-12 (Ecochemie) equipped with a frequency response analyzer. As counter and reference electrodes a platinum wire and a saturated calomel electrode were used.

3.1.3. References

[1] C. Deslouis, T. El Moustafid, M. Musiani, B. Tribollet, Electrochim. Acta **1996**,

41, 1343-1349.
[2] X. Ren, P. G. Pickup, J. Electroanal. Chem. **1997**, 420, 251-257.
[3] Q. Hao, V. Kulikov, V. M. Mirsky, Sens. Actuators B **2003**, 94, 352-357.
[4] V. Kulikov, V. M. Mirsky, T. L. Delaney, D. Donoval, A. W. Koch, O. S. Wolfbeis, Meas. Sci.Technol. **2005**, 16, 95-99.
[5] V. Kulikov, V. M. Mirsky, Meas. Sci.Technol. **2004**, 15, 49-54.
[6] J. Lippe, R. Holze, Synth.Met. **1991**, 43, 2927-2930.
[7] G. Zotti, Synth. Met. **1998**, 97, 267-272.
[8] W. Zhang, S. Dong, Electrochim. Acta **1993**, 38, 441-445.
[9] B. Feldman, P. Burgmayer, R. W. Murray, J.Am.Chem.Soc. **1985**, 107, 872-878.
[10] W. Albery, A. R. Mount, J.Chem.Soc., Faraday Trans. **1994**, 90, 1115-1119.
[11] A. Hallik, A. Alumaa, J. Tamm, V. Sammelselg, M. Vaartnou, A. Janes, E. Lust, Synth. Met. **2006**, 156, 488-494.

3.2. Characterisation of polythiophene in aqueous and organic solutions

Polythiophene films electrodeposited from common solvents, like acetonitrile or propylenecarbonate, usually show reduced chemical and physical properties, due to the high oxidation potential of the monomer, which leads to an overoxidation of the polymer during film growth.[1] Solvents like boron trifluoride diethylether (BFEE) lower the oxidation potential of thiophene, due to their Lewis acidity and as a result electropolymerisation of thiophene from this solvent results in polymer films of higher quality.[2]-[4] Such films were characterized earlier by cyclic voltammetry, FTIR and spectroelectrochemistry in organic media.[2]-[4] However, no data on the in-situ conductance of these films was reported and only a few publications describe the electrochemical behaviour of polythiophene in aqueous media.[5]-[7] However, these data are important for applications of this material in chemiresistors and electrochemical chemitransitors. Therefore we compared the electrochemical and spectroelectrochemical behaviour of these films in aqueous and organic solvents and investigated the influence of the oxidation state on the polymer-gold contact and bulk polymer conductivities

3.2.1. Results and Discussion

The potentiostatic electropolymerisation of thiophene from 90% BFEE / 10 % acetonitrile at 1.35 V vs. a Ag-wire yields smooth homogeneous films, which cover the gaps between the sensing strips completely and show a good stability and adhesion to the electrode (Fig. 13). At the high polymerization charge used for the coating of interdigitated electrodes this film has some surface inhomogenities (Fig. 13 A, B) while at the lower polymerization charge used for ITO coating for spectroelectrochemical measurements, an optically smooth coating was obtained (Fig. 13, C). Similar effects were observed by polymerization of poly-N-methlyaniline.[8]

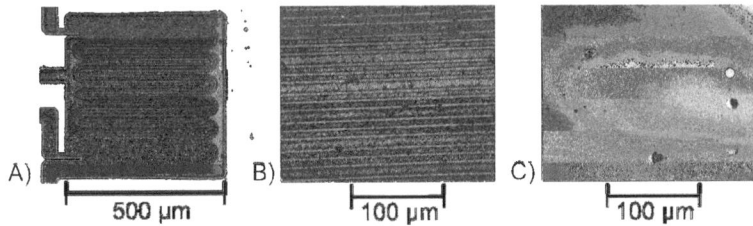

Figure 13. Interdigitated gold electrode coated by polythiophene by electropolymerization at +1.35 V from BFEE, polymerization charge 1.2 C/cm^2 (A). The same image under higher magnification (B) demonstrates some surface inhomogenities which are not present in the sample obtained on ITO surface at polymerization charge 25 mC/cm^2 (C).

After washing with acetonitrile and drying under nitrogen flow, the films were cycled in a 50 mM phosphate pH 2 buffer containing 0.1 M NaCl. The resulting voltammogram can be seen in Fig. 14 A. The anodic cycling potential was limited to +1.1 V. An application of a higher potential led to overoxidation and corresponding loss of electrochemical activity. A similar effect was described for various substituted polythiophenes.[5] It was found that the current response in aqueous solution was much lower than in acetonitrile (Fig. 14 B). This effect can be explained by the low penetration of highly hydrophilic ions from aqueous solution into the highly hydrophobic film of polythiophene.[9],[10] On the other hand the release of hydrophobic anions entrapped in polythiophene into the aqueous phase is also hindered.[11]

Results and Discussion

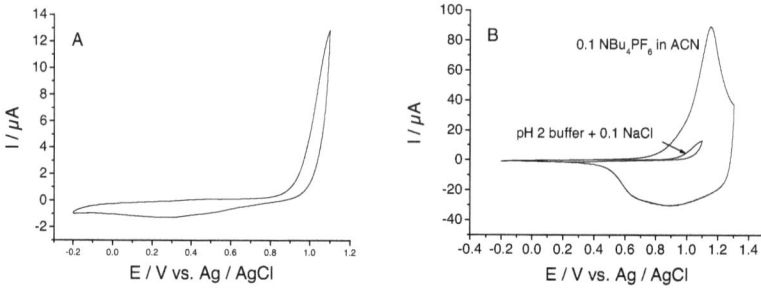

Figure 14. Cyclic voltammetry of polythiophene coated gold electrodes in aqueous solution (A and inner curve in B) of phosphate buffer (10 mM, pH 2) containing 0.1 M NaCl and in acetonitrile solution of 100 mM NBu$_4$PF$_6$ (B, outer curve). Sweep rate: 0.1 V/s.

Spectroelectrochemical measurements of Polythiophene-films deposited onto ITO-electrodes (Fig. 15) were consistent with the results obtained by cyclic voltammetry. The behaviour of these films in acetonitrile (Fig. 15, A) corresponds to the data reported in [4], while at the same potential scale in aqueous solution only slight optical changes were observed (Fig. 15, B). The results confirmed that polythiophene in its reduced form is more stable against oxidation in aqueous environment than in acetonitrile: the potential increase to 1.1 V leads to about 50% decrease of the concentration of reduced state of polythiophene in acetonitrile but to only about 12% decrease in aqueous environment.

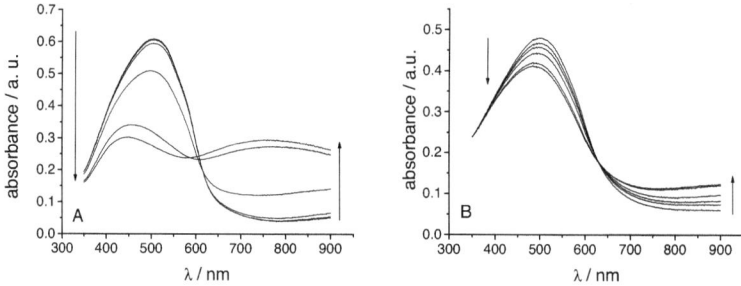

Figure 15. Influence of electrical potential on optical absorbance of polythiophene films on ITO electrodes in acetonitrile solution of 100 mM NBu$_4$PF$_6$ (A) and in aqueous solution of 10 mM phosphate buffer at pH 2 containing 0.1 M NaCl (b). The arrows on the figure indicate an increasing of the potential. The potential values recalculated to the potentials vs. Ag/AgCl (sat.) are -0.2 V, 0.4 V, 0.6 V, 0.8 V, 1.0 V and 1.1 V.

Results and Discussion

The results of the in-situ conductivity measurements (Fig. 16) at anodic potentials are consistent with the results obtained from cyclic voltammetry. It can be seen that the total change of conductivity in the aqueous solution is about three to four orders of magnitude but it increases up to over six orders of magnitude in the acetonitrile solution. In the reduced state the polythiophene film shows a higher conductance in the aqueous solution, however the conductance in acetonitrile is higher at potentials more than +0.8 V. The higher conductance in the reduced state of the polythiophene film in aqueous solution is probably due to the hindered reduction of the film in aqueous solutions. A suppression of the potential effect on polymer conductivity after polymer transfer from organic to aqueous medium was reported earlier for poly(3-methyl)thiophene.[6],[7] The difference in response was controlled by the anion used in the polymer synthesis.[7]

The potential influence on polythiophene resistance in aqueous media indicates clearly that the polymer has at least three potential dependent states. Some polymer oxidation at +0.25 V leads to increase of the polymer conductance for about one order of magnitude (Fig. 16). However, this oxidation is followed by very low changes of the polymer charge which can be hardly observed in voltammetric curve.

The potential influence on the bulk polymer resistance can be described quantitatively by simple the three-state model. It is suggested that there is an equilibrium between three states $A \leftrightarrow B \leftrightarrow C$ corresponding to three states of the polymer (reduced, slightly oxidized, strongly oxidized); the polymer units have charges n_x and molar conductivities g_x, where the index x indicates the state A, B or C. The behaviour of different units is assumed to be independent. The total conductivity of the polymer between electrodes is the sum of conductivites of all polymer states: $G = c_A\, g_A + c_B\, g_B + c_C\, g_C$, where c_x are the molar fraction of the corresponding species. Electrochemical potentials of all these states in equilibrium are equal, therefore: $\tilde{\mu}_A = \tilde{\mu}_B = \tilde{\mu}_C$. A substitution of $\tilde{\mu}_x = \mu_{0,x} + RT \ln a_x + n_x EF$ (here a_x is the activity, E is the electrical potential, and R, T, F have the usual meaning), and an approximation of activities by concentrations gives the equation:

$$G = \frac{g_C + \left[g_B + g_A \exp\left((E^0_{AB} - E)\frac{(n_B - n_A)F}{RT}\right)\right]\exp\left((E^0_{BC} - E)\frac{(n_C - n_B)F}{RT}\right)}{1 + \left[1 + \exp\left((E^0_{AB} - E)\frac{(n_B - n_A)F}{RT}\right)\right]\exp\left((E^0_{BC} - E)\frac{(n_C - n_B)F}{RT}\right)},$$

Results and Discussion

where E_{AB} and E_{BC} are $\dfrac{\mu_{0,A}-\mu_{0,B}}{(n_B-n_A)F}$ and $\dfrac{\mu_{0,B}-\mu_{0,C}}{(n_C-n_B)F}$ correspondingly and have the physical meaning of the redox potential of the reactions **A ↔ B** and **B ↔ C**. The two oxidized polymer states can correspond for example, to polaron and bipolaron states. Physically the model is similar to that suggested in [12] for the description of the spectroelectrochemical activity of polypyrrole.

The model was applied for fitting of the experimental results in aqueous solution (pH 2). Taking into account that in the reduced state the polymer is not charged, the value n_A was postulated to be zero. The values of n_B and n_C were 0.5 and 1.0. The fitting results are shown in the Fig. 16. A quantitative description of experimental data was obtained with oxidation potentials 330 mV and 1180 mV. Some electrochemical activity at these potentials is observed in voltammetry measurements (Fig. 14 A). The obtained ratio of molar conductivities is $g_A : g_B : g_C$ = 1:130:10000.

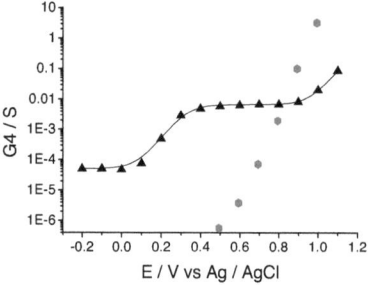

Figure 16. In-situ conductivity of polythiophene measured in acetonitrile solution of 100 mM NBu$_4$PF$_6$ (•) and in aqueous solution of 10 mM phosphate buffer containing 0.1 M NaCl at pH 2 (▲). The continuous line indicates a fitting of the experimental dependence by three-state model (see the text for details).

The kinetics of resistance changes induced by variation of electrode potential to anodic and to cathodic direction is shown in the Fig. 17. The results demonstrate that the oxidation in aqueous solution is faster than the reduction. The steady state resistance is about 13 kOhms, which is much lower than the resistance of the reduced film in acetonitrile. Therefore it is suggested that at least at thick films, used

Results and Discussion

for in-situ resistance measurements, the reduction in aqueous media at the used potentials is incomplete.

The kinetics of contact and bulk resistances on potential changes is quite different: contact resistance reaches fast its steady-state value in about 30 s (for polymer oxidation) or a couple of minutes (for polymer reduction) while relaxation of bulk polymer resistance is much slower process. This indicates clearly that the values of contact and polymer resistance obtained by analysis of simultaneously measured two- and four-point resistances really correspond to different physico-chemical processes. This is the first kinetic measurement of the metal/polymer contact resistance.

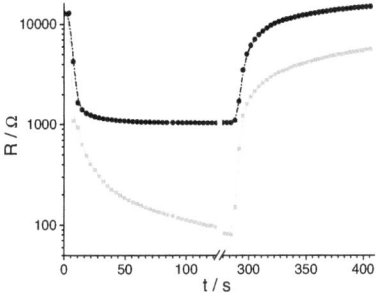

Figure 17. Kinetics of bulk (■) and contact (●) film resistances by potential changes from -0.2 V to 1.1 V and back in aqueous solution of 10 mM phosphate buffer containing 0.1 M NaCl at pH 2

Fig. 18 shows the dependence of the contact resistance on the potential in comparison with polymer resistances measured in two- and four-point configurations in aqueous environment and in acetonitrile. It is interesting that while the contact resistance between 0.4 V and 0.8 V drops, the polymer resistance remains almost constant. Oppositely, at potentials higher than 0.9 V the contact resistance is almost constant or displays a small increase while the polymer resistance decreases strongly. In acetonitrile the behaviour of the contact resistance is very similar to that of the polymer resistance, but like in aqueous solution it is almost constant at potentials higher than 0.9 V. The behaviour of the polymer resistance at high potential is very close to that in aqueous solution.

39

Results and Discussion

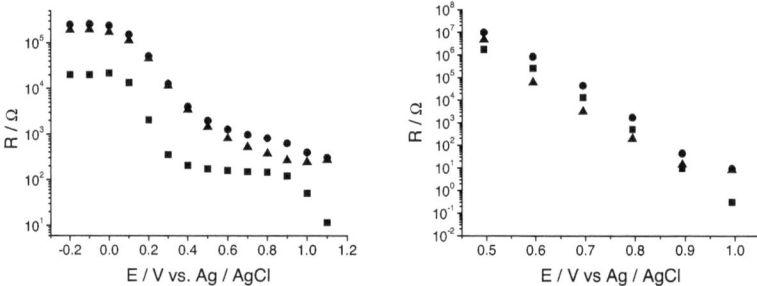

Figure 18. Potential influence on the resistances of polythiophene films in aqueous solution (10 mM phosphate, pH 2, 0.1 M NaCl) (left) and in acetonitrile (0.1 M NBu$_4$PF$_6$) (right) measured by 2(●)- and 4(■)-point configuration and the contact resistance(▲) calculated according to (3).

3.2.2. Experimental

Polythiophene films were prepared at a constant potential of 1.35 V vs. a silver wire from a 0.05 M solution of thiophene (Merck) in 90% BFEE (Sigma) 10% acetonitrile (Baker, HPLC grade). The oxidation charge was about 1200 mC/cm^2 for conductivity measurements and 25 mC/cm^2 for the spectroelectrochemical measurements. Cyclic voltammetry was performed on a General Purpose Electrochemical System Autolab PGSTAT-12 (EcoChemie). A one compartment cell with three electrodes, working electrode, platinum wire counter electrode and either silver chloride electrode (Ag/AgCl, sat. KCl in water) for aqueous solutions or a silver nitrate electrode (Ag/AgNO$_3$, 10mM AgNO$_3$, 0.1 M tetrabutylammonium hexafluorophosphate (NBu$_4$PF$_6$, Fluka, electrochemical grade) in acetonitrile) for acetonitrile solutions was used. The potential of the silver nitrate electrode against the silver chloride reference electrode was +0.344 mV. All potentials obtained in organic solutions were recalculated to the silver chloride reference. Measurements in aqueous media were done in 10 mM pH 2 phosphate (Merck) buffer containing 0.1 M NaCl (Merck). Deionised water additionally purified by Millipore Milli-Q system was used. For measurements in acetonitrile a solution of 0.1 M tetrabutylammonium hexafluorophosphate was used. All measurements were performed at room temperature. In-situ conductivity measurements were performed using simultaneous two- and four point measurement– technique. The polymer potential was controlled versus the reference electrodes described above. The same reference electrodes were used also for spectroelectrochemical measurements. These measurements were performed in two – electrode configuration using a Cary 50Bio spectrophotometer from Varian.

3.2.3. References

[1] G. Wallace, G. Spinks, L. Kane-Maguire, P. Teasdale, Conductive Electroactive Polymers: Intelligent Materials Systems, CRC **2003**.
[2] X. Li, Y. Li, Journal of Applied Polymer Science **2003**, 90, 940-946.
[3] G. Q. Shi, C. Li, Y. Q. Liang, Adv. Mater. **1999**, 11, 1145-1146.
[4] C. Alkan, Adv. Funct. Mater. **2003**, 13, 331-336.
[5] S. Sunde, G. Hagen, R. Oedegaard, Synth. Met. **1991**, 43, 2983-2986.
[6] S. Sunde, G. Hagen, R. Odegard, J. Electrochem. Soc. **1991**, 138, 2561-2566.
[7] E. Lankinen, G. Sundholm, P. Talonen, T. Laitinen, T. Saario, J. Electroanal. Chem. **1998**, 447, 135-145.
[8] Q. Hao, M. Rahm, D. Weiss, V. M. Mirsky, Microchim. Acta **2003**, 143, 147-153.
[9] Z. Zhang, L. Qu, G. Shi, J. Mater. Chem. **2003**, 13, 2858-2860.
[10] A. Fikus, U. Rammelt, W. Plieth, Electrochim. Acta **1999**, 44, 2025-2035.
[11] S. A. Refaey, G. Schwitzgebel, New Polym. Mater. **1995**, 4, 301-308.
[12] T. Amemiya, K. Hashimoto, A. Fujishima, K. Itoh, J. Electrochem. Soc. **1991**, 138, 2845-2850.

Results and Discussion

3.3. Six electrode electrochemical transistor

3.3.1. Six electrode measurements

Electrochemcial transistors provide the possibility to control the oxidation state of conducting polymers, which is important in conducting polymer based conductometric chemo- and biosensors. In such sensors which respond to pH,[1],[2] ions[3] or analyte binding on receptors linked to the conducting polymer backbone,[4] the electrochemical control of the redox state provides a higher reliability and reproducibility of the sensor response as changes in the redox potential can be excluded as origin of the response. Furthermore in sensors which measure changes in redox potential the electrochemical control provides a fast regeneration and conditioning of the sensor before the measurement.[5]-[7]

For such applications it is very convenient to have the gate electrode (reference and counter electrode) on one sensor chip together with the electrodes for resistance measurements. For this purpose the electrodes described above in chapter 2.2 were designed. Here we will test the performance of such six electrode electrochemical transistor composed of a Ag or Ag / AgCl modified reference electrode, polyaniline or polythiophene modified resistance measurement electrodes (working electrode) and polyaniline or polythiophene modified counter electrodes.

3.3.1.1. Results and Discussion

The reference electrodes were obtained by galvanostatic deposition of silver from an aqueous solution of 10 mM $AgNO_3$, 20 mM EDTA, 120mM NH_3 and 80mM NaOH on the reference electrode gold strip at 10 μA for 1 h. In case of Ag / AgCl electrodes the AgCl coating was obtained by dipping the electrode in a solution of 100 mg / ml $FeCl_3$ in ethanol for 5 minutes.[8] To remove any unstable Ag or AgCl the modified electrode was treated by ultrasound in water for several minutes. The resulting Ag or Ag / AgCl layers were homogeneous and very adherent to the gold electrode (Fig. 19). After reference electrode modification conducting polymer films were deposited on the working and counter electrodes (see Experimental part).

Results and Discussion

Figure 19. Modified electrodes for the use in electrochemical transistors. Ag modified reference electrode (left) without and with polythiophene modified working and counter electrode. Ag / AgCl modified reference electrode (right) without and with polyaniline modified working and counter electrode.

To test their performance and to calibrate the reference electrodes, the redox behaviour of $K_3Fe(CN)_6$ and ferrocene on the gold working electrodes was investigated by cyclic voltammetry using the internal on-chip reference electrode and compared to the behaviour obtained by using a external reference electrode . Fig. 20 shows the voltammograms of $K_3Fe(CN)_6$ and ferrocence measured on interdigitated gold electrodes with an external and internal on-chip reference electrodes. The $K_3Fe(CN)_6$ peak shift of the on-chip Ag / AgCl reference electrode to the external Ag / AgCl (sat. KCl) reference electrode in 0.1 M HCl is 87 mV which is close to the theoretically predicted value of 89 mV. During long time cycling (60 scans) the on-chip reference electrode also showed a good potential stability. The pseudo silver reference electrode was used for measurement in non-aqueous solutions. From the ferrocence peaks measured in acetonitrile with NBu_4PF_6 as electrolyte a potential difference of 65 mV versus an Ag / Ag$^+$ (10mM, ACN) reference electrode was measured.

Results and Discussion

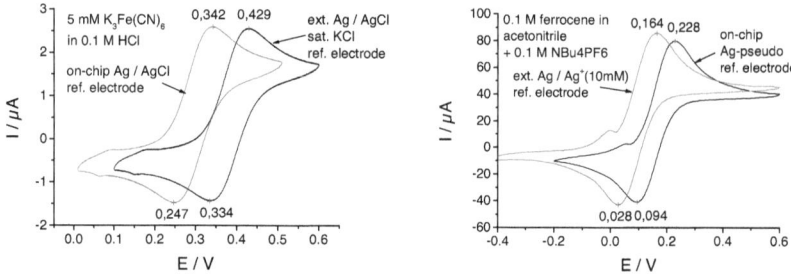

Figure 20. Calibration of the Ag / AgCl (A) and Ag (pseudo-) (B) reference electrodes by cyclic voltammetry of $K_3Fe(CN)_6$ in 0.1 M HCl (A) or ferrocene in acetonitrile containing 0.1 M NBu_4PF_6. Scan rate: 0.1 V / s

To demonstrate the possibility of an electrochemical control of a conducting polymer layer on the four resistance measurement electrodes by using the on-chip reference and counter electrodes, a thin polyaniline layer was deposited over the four electrodes for resistance measurements. Additionally the counter electrode was modified by polyaniline. The polyaniline film on the counter electrode serves only as charge reservoir to balance the charge created on the working electrode during polyaniline oxidation / reduction. Figure 21 A shows the two and four point resistance of the described electrochemical transistor measured during potential cycling between -0.08 V and 0.8 V (Scan rate 10 mV / s) in 1 M HCl.

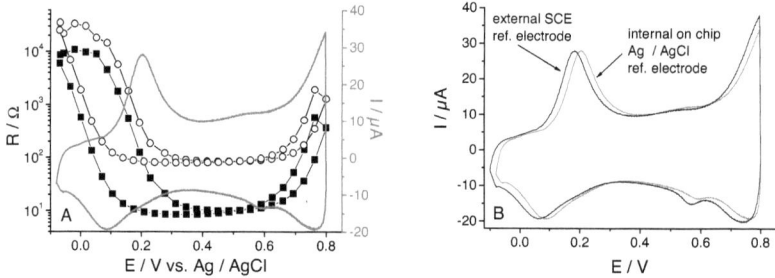

Figure 21. Potential dependent two-(○) and four point (■) resistance of polyaniline in 1 M HCl measured in the six-electrode electrochemical transistor configuration described in the text. Scan rate: 10 mV / s (A), Cyclic voltammogram of the polyaniline film in 1 M HCl measured by using the on-chip counter and reference electrodes (A and B) and external counter and reference electrodes (B). Scan rate: 0.1 V / s.

Results and Discussion

As described earlier,[9],[10] upon oxidation of the leucoemeraldine form of polyaniline to the emeraldine salt form the resistance drops about 4 orders of magnitude and increases again by the second oxidation to the pernigraniline form. The voltammogram of polyaniline was recorded without resistance measurements at a scan of 0.1 V / s and is almost identical to the voltamogram measured by using external reference and counter electrodes (Fig. 21 B).

Both methods, simultaneous two- and four-point measurements and cyclic voltammetry, demonstrate the possibility to electrochemically control a conducting polymer film on the electrodes for resistance measurements by using on-chip reference and counter electrodes.

Besides polyaniline and an aqueous electrolyte the described electrochemical transistor was also tested by using polythiophene and a gel electrolyte based on poly(2-acrylamido-2-methyl-1-propane-sulfonic acid) (PAMPSA). Figure 22 shows the potential dependent two- and four-point resistance of this electrochemical transistor. By deposition of a gel electrolyte over the sensor it is possible to use this electrochemical transistor also in the gas phase as chemical sensor. A more detailed characterisation of this electrochemcical transistor and an application as gas sensors is demonstrated in the next chapter.

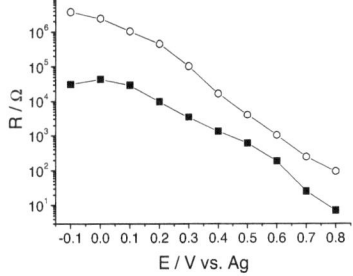

Figure 22. Potential dependent two-(○) and four point (■) resistance of polythiophene coated by a PAMPSA electrolyte containing LiClO$_4$, measured in the six-electrode electrochemical transistor configuration described in the text by changing the potential from -0.1 V to 0.8 V in 0.1 V steps.

Results and Discussion

3.3.2. Electrochemical regeneration of conducting polymer based gas sensors

Conducting polymers[11], carbon nanotubes[12] and graphenes[13] show strong changes in resistance upon interaction with oxidizing and reducing gases[14] and are therefore perspective materials for such sensors. Advantages of these materials include a high sensitivity, the ability to work at room temperature, the compatibility to organic flexible electronics and a low price. However the sensor recovery after exposure to analytes is very slow.[15] Therefore, it is important to find a possibility to speed up this process. Several approaches like heating[16] or UV-light exposure[17] were suggested. However these methods are unpractical, poor-compatible with flexible organic electronics and require exact balancing of temperature extension of all contacting materials. As described above in liquid electrolytes electrochemical regeneration is used in the detection of redox active analytes. By replacing the liquid electrolyte with a gel electrolyte electrochemical control of the sensor film is also possible in the gas phase (Figure 22). This opens the possibility of an electrochemical regeneration of conducting polymer based gas sensors. A application of electrochemical transistors in the gas phase was already tested for humidity sensors,[18],[19] but no application of an electrochemical regeneration of conducting polymer based gas sensors was described yet.

3.3.2.1. Results and Discussion

The design and wiring of the solid state electrochemical transistor for an application as gas sensor are shown in Figure 23. The sensor was realized by electropolymerisation of a thin polythiophene film from a 0.2 M solution of thiophene in a mixture of 90 % boron trifluordiethyletherate (BFEE) and 10 % acetonitrile on the four resistance measurement electrodes and the counter electrode like shown in Figure 19. After the modification of the electrodes the chip was covered with a thin film of a poly(2-acrylamido-2-methyl-1-propane-sulfonic acid) (PAMPSA) electrolyte containing $LiClO_4$. To obtain a polymer gel electrolyte with a sufficient high conductivity at normal humidity ethylenglycol and sorbitol were added to the polymer solution in water.[20]

Results and Discussion

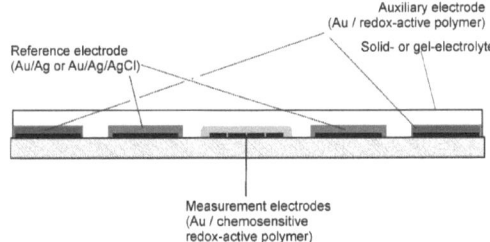

Figure 23. Design of the electrochemical transistor for applications as gas sensor.

At a sufficient (30 %) humidity the PAMPSA electrolyte provides enough conductivity to allow the electrochemical control of the conducting polymer film deposited on the working electrodes. Figure 24 demonstrates the cyclic voltammetric curve of polythiophene on the sensor in combination with the dependence of the drain current on the gate voltage. The stable voltammetric curve demonstrates the possibility to control the redox state of the polymer deposited on the measurement electrodes by reference and counter electrodes contacting this polymer through the polymer gel-electrolyte. Upon switching the polymer from its reduced state at -0.1 V to its oxidized state at 0.8 V the conductance changed for over four orders of magnitude. Figure 24 B shows the drain current voltage characteristics of the sensor measured in the oxidized state at a gate potential of 0.7 V, in some intermediate oxidation state at 0.4 V and in the reduced state at -0.1 V. In all cases the characteristic was linear from 1 mV to 50 mV.

Both methods cyclic voltammetry and in-situ conductance measurements prove that the oxidation state of the conducting polymer film can be controlled in the solid state electrochemical transistor. This opens up the possibility for an electrochemical regeneration of the conducting polymer sensor film after exposure to redox active gases.

Results and Discussion

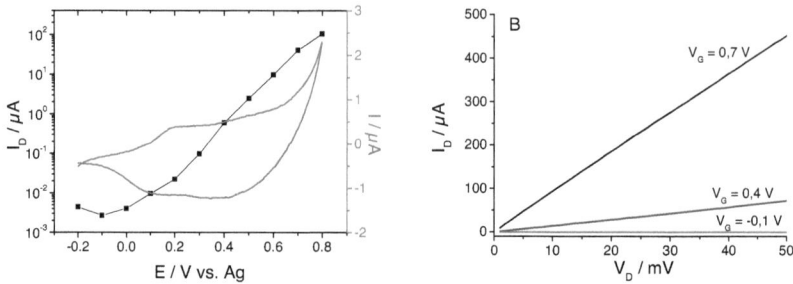

Figure 24. Cyclic voltammogram of the polythiophene modified sensor covered by a PAMPSA electrolyte containing $LiClO_4$ at 10 mV / s versus a silver pseudo reference electrode in combination with the gate voltage dependent drain current at a drain voltage of 10 mV (A) and drain current-voltage characteristics of the electrochemical chemotransistor at different gate voltages (B).

As an example the difference in sensing performance between a usual polythiophene chemoresistor and our electrochemical transistor is illustrated in Figure 25. Upon exposure to NO_2 the polythiophene film is oxidized, leading to an increase in its conductivity and a corresponding increase in the drain current. Both sensors exhibited a high sensitivity towards nitrogen dioxide. However the recovery of the chemoresistor performed by perfusion with synthetic air was not complete even after 1 hour (Fig. 25 B). The sensor based on electrochemical transistor (Fig. 25 A) was electrically regenerated within a few minutes.

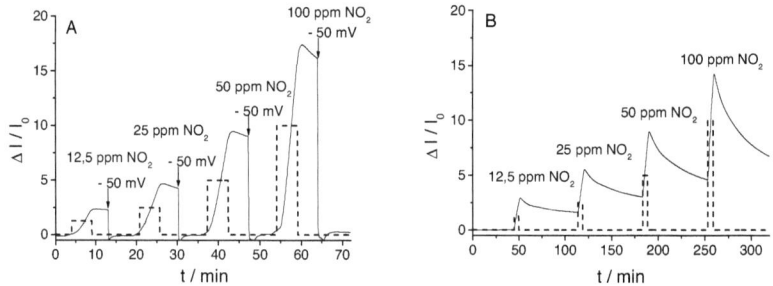

Figure 25. Current response to different concentrations of NO_2 in synthetic air of the electrochemical transistor (A) and a chemiresistor (B). The dashed line represents the NO_2 pulses. The pulse duration was 5 minutes. Regeneration time was 10 minutes in the case of the electrochemical transistor and 65 minutes in the case of the chemoresistor.

Figure 26 shows concentration dependence of the sensor signal. The detection limit of the sensor calculated from the signal to noise ratio is 60 ppb.

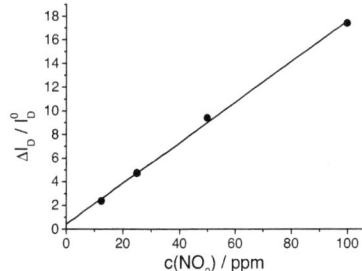

Figure 26. Concentration dependent response of the polythiophene based electrochemical transistor NO_2 sensor

The regeneration procedure suggested can be easy implemented into sensor devices. The suggested type of integrated electrochemical transistor can be applied also to other types of chemical sensors and biosensors exploiting conducting and red-ox active materials as receptors and transducers, including carbon nanotubes and graphenes.

3.3.3. Experimental

Linear gold electrodes were used for the design of the electrochemical transistor. The reference electrode was made by galvanostatic deposition of silver from a solution of 10 mM $AgNO_3$, 20 mM EDTA, 120mM NH_3 and 80mM NaOH on the reference electrode gold strip at 10 µA for 1 h. As counter electrode a platinum electrode and as reference electrode a Cu / $CuSO_4$ (sat.) electrode was used. In case of Ag / AgCl electrodes the AgCl coating was obtained by dipping the electrode in a solution of 100 mg / ml $FeCl_3$ in ethanol for 5 minutes.[8] Polyaniline (PANI) based electrochemical chemotransitors were obtained by electropolymerisation of aniline from a 0.05 M solution of aniline in 1 M HCl at a potential of 0.8 V vs. Ag / AgCl (sat.). The polymerization charge was 1 mC for the working electrodes and 3 mC for the counter electrode. Polythiophene was electropolymerized from a 0.2 M solution of thiophene in 90 % boron trifluoride diethyl etherate (BFEE) / 10 % acetonitrile at 1.3 V vs. Ag / Ag$^+$ (10mM) with a charge of 2.5 mC on the working electrodes and 7.5 mC on the counter electrode. This electrolyte was chosen as BFEE lowers the oxidation potential of thiophene, allowing one to avoid overoxidation of polythiophene during the polymerisation process.[21] After modification of the electrodes the chip was covered by a thin layer of a poly(2-acrylamido-2-methyl-1-propane-sulfonic acid) (PAMPSA) electrolyte. The electrolyte was prepared by mixing 500 mg of a 15 % solution of PAMPSA in water, 75 mg ethyleneglycol, 25 mg sorbitol, 16 mg LiClO$_4$ and 0.5 mL water. Ethylenglycol and sorbitol were added to the polymer solution in water to obtain a polymer gel electrolyte with a sufficient high conductivity at normal humidity.[20] The electrolyte layer was allowed to dry on air for several hours before measurements.

A potentiostat Radiometer-PGP201 was used for potential control. The characterisation of the electrochemical transistor gas sensor without analyte was performed in air over a saturated CaCl$_2$ providing constant humidity of about 30%. The sensor response was measured in flow through cell at flow rate of 120 ml / min. A 300 ppm NO$_2$ in nitrogen was diluted with synthetic air to the corresponding concentrations using a home-made gas mixing device with computer control. To humidify the synthetic air, it was passed through the headspace of a water containing flask before entering the measurement flow cell.

During NO$_2$ exposure the measurement electrodes were disconnected from the potentiostat. For regeneration of the electrochemical transistor it was first flushed with synthetic air for further 5 min, reconnected to the potentiostat and then a gate potential of -50 mV was applied for about 30 s. To achieve a fast regeneration, the reduction was carried out until the current decreased to about 50 % of the initial baseline drain current; further disconnecting of the gate leads to increase of the drain current till its initial value.

3.3.4. References

[1] M. Nishizawa, T. Matsue, I. Uchida, Sens. Actuators B **1993**, 13, 53-56.

[2] M. Nishizawa, T. Matsue, I. Uchida, Anal. Chem. **1992**, 64, 2642-2644.

[3] V. Saxena, V. Shirodkar, R. Prakash, Appl. Biochem. Biotechnol. **2001**, 96, 63-

69.
[4] B. Fabre, L. Taillebois, Chem. Comm. **2003**, 2003, 2982-2983.
[5] P. N. Bartlett, P. R. Birkin, Anal. Chem. **1994**, 66, 1552-1559.
[6] P. N. Bartlett, J. H. Wang, E. N. K. Wallace, Chem. Commun. **1996**.
[7] P. Bartlett, Y. Astier, Chem. Comm. **2000**, 2000, 105-112.
[8] O. Segut, B. Lakard, G. Herlem, J. Rauch, J. Jeannot, L. Robert, B. Fahys, Anal. Chim. Acta **2007**, 597, 313-321.
[9] E. W. Paul, A. J. Ricco, M. S. Wrighton, J.Phys.Chem. **1985**, 89, 1441-1447.
[10] W. Huang, B. Humphrey, A. MacDiarmid, J. Chem. Soc., Faraday Trans. 1 **1986**, 82, 2385-2400.
[11] J. Janata, M. Josowicz, Nat. Mater. **2003**, 2, 19-24.
[12] D. Kauffman, A. Star, Angew. Chem. Intern. Ed. **2008**, 47, 6550-6570.
[13] F. Schedin, A. K. Geim, S. V. Morozov, E. W. Hill, P. Blake, M. I. Katsnelson, K. S. Novoselov, Nat. Mater. **2007**, 6, 652-655.
[14] U. Lange, N. Roznyatovskaya, V. Mirsky, Anal. Chim. Acta **2008**, 614, 1-26.
[15] H. Bai, G. Shi, Sensors **2007**, 7, 267-307.
[16] J. Kong, N. R. Franklin, C. Zhou, M. G. Chapline, S. Peng, K. Cho, H. Dai, Science **2000**, 287, 622-625.
[17] V. Dua, S. P. Surwade, S. Ammu, X. Zhang, S. Jain, S. K. Manohar, Macromolecules **2009**, 42, 5414-5415.
[18] S. Chao, M. Wrighton, J.Am.Chem.Soc. **1987**, 109, 6627-6631.
[19] D. Nilsson, T. Kugler, P. O. Svensson, M. Berggren, Sens. Actuators B **2002**, 86, 193-197.
[20] M. Hamedi, R. Forchheimer, O. Inganas, Nat. Mater. **2007**, 6, 357-362.
[21] G. Q. Shi, C. Li, Y. Q. Liang, Adv. Mater. **1999**, 11, 1145-1146.

Results and Discussion

3.4. Electrochemical transistors with ion selective gate electrodes

In electrochemical transistor based sensors not only the conducting film bridging the source and drain electrodes can be used as chemosensitive element, but similar to ion-sensitive field effect transistors also the gate(reference) electrode.[1]-[4] In this case the conducting film works only as transducer element. This principle has been used for detecting redox couples like H^+/ H_2,[1] p-aminophenol[2] and hydrogen peroxide.[3],[4]
Ion selective electrochemical transistors have been reported, using an ion-selective membrane on top of the conducting film or by using a conducting film which changes its conductivity in the presence of specific ions.[5]-[7] However there are no reports about using an ion-selective electrode as gate electrode in electrochemical transistors. In this chapter this combination is evaluated by using a K^+-selective electrode as gate electrode in a PEDOT / PSS based electrochemical transistor.

3.4.1. Results and Discussion

The K^+-selective electrode was made by dipping a platinum wire first in a solution of PEDOT / PSS to obtain a very thin layer of this conducting polymer composite on the wire, which should work as ion-to electron transducer[8] and subsequently coating the so modified wire by a plasticized PVC membrane containing valinomycin as K^+-selective ionophor. After preconditioning this electrode in a 0.01 M KCl solution, the potentiometric response of this electrode towards K^+ was measured in 0.01 M NaCl with increasing concentrations of KCl versus a saturated calomel electrode (Fig. 28 B). An increase in the potential by 40 mV / decade was measured by increasing the concentration from 10^{-5} M to 10^{-1} M K^+. After this the electrode was connected as reference electrode together with a platinum counter electrode to a PEDOT PSS covered interdigitated electrode for resistance measurements in an electrochemical transistor configuration. The potential dependent conductance of the PEDOT PSS film versus a saturated calomel electrode is shown in Figure 27.

Results and Discussion

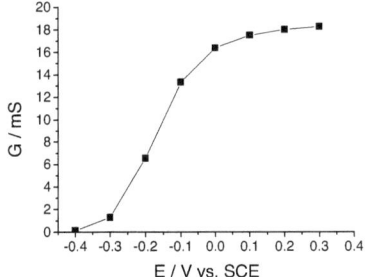

Figure 27. Potential dependent conductance of PEDOT PSS in 10 mM NaCl

The potential between the K⁺-selective electrode and the PEDOT PSS covered electrode was set to – 150 mV, which is about -300 mV vs. a saturated calomel electrode. Increasing concentrations of K⁺-ions in solution lead to a positive shift of the potential of the K⁺-selective electrode (Fig. 28 B), which induces changes in the oxidation state and therefore the conductance of the PEDOT PSS film. (Fig. 28 A).

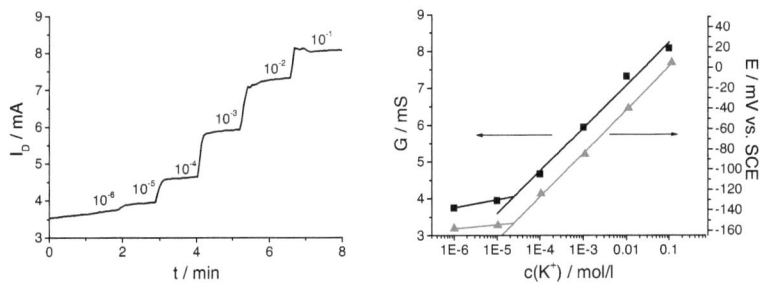

Figure 28. Change of the drain current upon increasing the K⁺ concentration to the values given in the figure in mol / l in a 10 mM solution of NaCl (A) and calibration plot of the electrochemical transistor based sensor and the potentiometric sensor using the same ion-selective electrode.

To test whether the addition of KCl to the 0.01 M NaCl solution has an direct influence on the conductivity of the PEDOT PSS layer, the experiment was repeated with a saturated calomel reference electrode at -280 mV instead of the K⁺-selective electrode. Figure 29 A shows that at addition of concentrations higher than 1 mM KCl the conductance of the PEDOT PSS layer decreased. This is in contrast to the

53

Results and Discussion

measured increase in conductance using the K⁺-selective electrode as reference electrode (Fig. 28 A). An explanation of this behaviour is an increased reduction of PEDOT PSS due to the higher ionic strength. The redox equilibrium of conducting polymers is influenced not only by the redox potential of the conducting polymer, but also by the ion flux rate in and out of the polymer (see e.g. chaper 3.2). By correcting the response of the K⁺-selective electrochemical transistor presented in Figure 27 by the response of PEDOT PSS itself on the addition of KCl an improved linearity was obtained in the calibration curve (Fig. 29 B), which is in accordance with the linearity of the potentiometric response of the K⁺-selective electrode (Fig. 28 B).

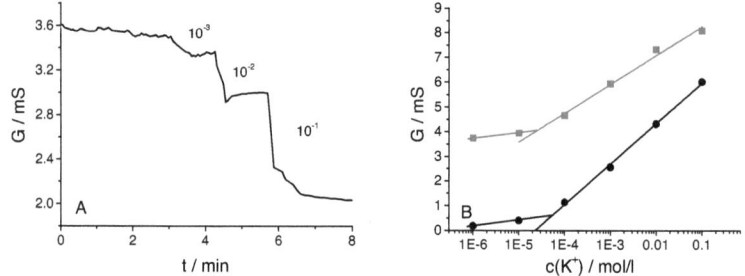

Figure 29. Influence of the addition of KCl to a 0.01 M solution of NaCl on the conductivity of a PEDOT PSS film at -300 mV vs. SCE (A). Calibration curve of the K⁺-selective electrode after correction of the influence of KCl addition on the conductivity of PEDOT PSS (●) in comparison to the uncorrected curve (■) (B).

3.4.2. Experimental

The K⁺-selective electrode was made by precoating a platinum wire with a commercially available PEDOT PSS solution (Sigma Aldrich, conducting grade) by simple dipping it several times in the solution and drying on air. After this the electrode was dip-coated by a solution of 270 mg polyvinlychloride in 3 ml tetrahydrofurane containing additionally 520 µL dibutyl sebacate as plasticiser and 50 µL of a valinomycin solution (80 mg / ml in methanol). After drying the electrode, it was preconditioned in 0.01 M KCl for 12 h. The electrode for conductance measurements was made by dropcoating 2 µL of a solution of 250 µL of the PEDOT PSS solution in 10 ml water containing 1 % (v/v) dimethylsulfoxide on the sensor spot of the interdigitated electrode on a hotplate at about 80°C.

Measurements of the electrochemical transistor were made using a three electrode configuration with either a saturated calomel electrode or the K^+-selective electrode as reference electrode, a platinum wire as counter electrode and the electrode for resistance measurements as working electrode. Resistance measurements were performed as described in chapter 2 using a pulse voltage of 50 mV.

Potentiometric measurements of the K^+-selective electrode were performed using a high impedance voltmeter (Keithley 175) in combination with a saturated calomel electrode.

3.4.3. References

[1] E. W. Paul, A. J. Ricco, M. S. Wrighton, J.Phys.Chem. **1985**, 89, 1441-1447.
[2] Y. Astier, P. Bartlett, Bioelectrochem. **2004**, 64, 15-22.
[3] D. J. Macaya, M. Nikolou, S. Takamatsu, J. T. Mabeck, R. M. Owens, G. G. Malliaras, Sens. Actuators B **2007**, 123, 374-378.
[4] Z. Zhu, J. T. Mabeck, C. Zhu, N. C. Cady, C. A. Batt, G. G. Malliaras, Chem. Commun. Y1 – 2004,1556-1557.
[5] Z. Mousavi, A. Ekholm, J. Bobacka, A. Ivaska, Electroanal. **2009**, 21, 472-479.
[6] D. A. Bernards, G. G. Malliaras, G. E. Toombes, S. M. Gruner, Appl. Phys. Lett. **2006**, 89, 053505-3.
[7] V. Saxena, V. Shirodkar, R. Prakash, Appl. Biochem. Biotechn. **2001**, 96, 63-69.
[8] A. I. Johan Bobacka, Electroanal. **2003**, 15, 366-374.

Results and Discussion

3.5. Polyaniline metal nanoparticle layer by layer composites

The layer-by-layer (LbL) deposition based on interactions between polyions with alternating charges first introduced by Decher et al.[1] offers a simple procedure for formation of multilayered structures. The approach has been also explored for the deposition of different conducting polymers, including polyaniline (PANI),[2],[3] which belongs to the most studied conducting polymers in these investigations. Various species, i.e. polyanions,[3],[4] polycations,[5]-[7] semiconductor particles,[8]-[11] non-ionic soluble polymers,[12] and more recently - metal complexes,[13]-[15] proteins,[16]-[18] enzymes[19] and carbon nanotubes[20] were used in combination with PANI, sulfonated PANI or other polyaniline derivatives. Recently also carboxylic acid modified gold nanoparticles were used for LbL deposition in combination with PANI.[21],[22]

Metal nanoparticle modified electrodes were used in chemical sensors mainly due to the electrocatalytical properties of the nanoparticles or due to the possibility to immobilize (bio)receptors on the nanoparticle surface (for details see chapter 1.3.1.).[23]

3.5.1. Polyaniline gold nanoparticle composite

PANI gold nanoparticle (AuNP) composites have been formed by electrochemical deposition of gold in PANI,[24] by reduction of tetrachloroaurate[24],[25] or tetrabromoaurate[26] by PANI. Recently, the LbL technique was applied to form composites of PANI and anion-coated AuNPs.[21],[22] The incorporation of gold nanoparticles in PANI leads to new chemosensitive materials.[27]

In this chapter the synthesis and characterization of a multilayered PANI – AuNP nanocomposite using gold nanoparticles synthesized by the Turkevich method[28] is described. Further an approach to use the incooperated nanoparticles as receptors for sulfuric compounds and mercury is tested. A high affinity of gold to mercury(0) and sulfuric compounds is well known and is used actively for the development of chemosensors.[29],[30] In the case of a close contact of PANI and AuNPs, the interaction of gold with these compounds may affect the electrochemical and conductive properties of the PANI – AuNPs nanocomposite. To achieve this

Results and Discussion

interaction, the nanoparticles should be either uncoated or coated by a weakly adsorbed monomolecular layer. This was the reason to focus on a composite material prepared by using gold nanoparticles without further modification or with just a dialysis treatment.

3.5.1.1. Results and discussion

AuNPs-PANI coatings consisting of three bi-layers were used for TEM imaging (Fig. 1). The size of the gold nanoparticles immobilized in the LbL structure varies between 15 – 30 nm. The particles density is $4.2 \cdot 10^{10}$ cm^{-2}. The AuNPs are well dispersed into the polymer phase and even in the case of agglomerates they visibly do not contact directly each other. Individual particles are coated by a shell, which is not the case for AuNPs directly cast on a TEM mesh without PANI. Thus the LbL deposition technique results in the formation of an adsorbed protective PANI layer which prevents further agglomeration of AuNPs.

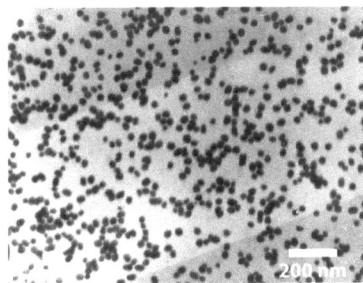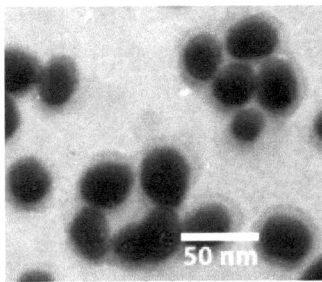

Figure 30. TEM images of a three bi-layer nanocomposite consisting of AuNPs and PANI at different magnifications.

An increase in the visible absorbance range was found by monitoring the subsequent adsorption of four PANI-AuNPs bi-layers using UV-Vis spectroscopy (Fig. 31 A). These measurements were carried out in air after rinsing the LbL deposited PANI-AuNPs structures with diluted HCl solution. The absorption spectra are composed of at least two absorbance bands the first one being the gold plasmonic band with the typical absorption maximum at about 520 nm,[31] which overlaps the second absorbance band with a maximum at about 625 nm. A comparison between the

Results and Discussion

spectra of AuNPs in solution, LbL assembled PANI – AuNPs and electropolymerized PANI-citrate is presented in Fig. 31 B. It is evident that the second absorption peak observed in the PANI – AuNPs case could be attributed neither to non-interacting AuNPs (a sharp single maximum at about 520 nm is observed in this case) nor to doped PANI - a large absorbance band starts emerging for much longer wavelengths. The deconvolution of the absorbance peaks (Fig. 31 A) shows that the ratio of the absorbance within the two observed bands is changing with number of deposited bilayers, the second absorbance band becoming dominant after the deposition of the second bilayer. On the other hand the deconvoluted short wavelength absorption peak shifts gradually from 520 nm (equivalent to the absorption of AuNPs dissolved in solution) to about 550 nm. A bathochromic shift in the AuNPs gold palsmonic band can result from the variation of the refractive index of the AuNPs environment.[32] A gradual shift in the gold plasmonic peak was already observed when increasing the number of polyelectrolyte layers deposited over an initially adsorbed AuNPs.[33]

The appearance and growth of the second long wavelength absorbance band could be possibly attributed to a collective resonance of the AuNPs with increasing density in the course of the LbL deposition. Double band absorbance spectra have been already found in the LbL systems AuNPs - poly (ethylene imine)[33] and AuNPs-poly(allylaminehydrochloride)-PSS.[34] It was suggested that the collective surface plasmon band depends strongly on the interparticle separation.[34]

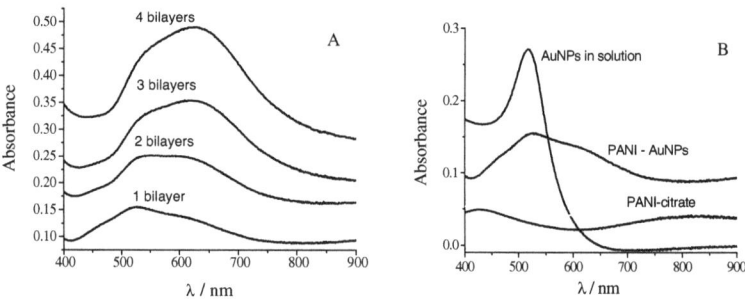

Figure 31. Absorption spectra obtained for deposition of four subsequent bi – layers of PANI – AuNPs (a) and a comparison between absorbance spectra of PANI – AuNPs, PANI – PSS and freshly synthesized AuNPs. The measurements of PANI – AuNPs and PANI – PSS nanocomposites were perfomed in air.

Results and Discussion

The electrochromic properties of the PANI – AuNPs nanocomposite were studied by in-situ spectroelectrochemical measurements performed on an ITO electrode covered with four PANI-AuNPs bilayers. The measurements were carried out in acidic solution (0.5 M H_2SO_4) which means that PANI was additionally doped and thus in a higher oxidation state. The absorbance spectra (Fig. 32) show the appearance of a single band (around 550 nm) that corresponds to the position of the deconvoluted gold plasmonic peak for the 4-bilayers structure (Fig. 31 A). The second absorbance band disappears from the absorbance spectra thus indicating disturbance of the collective resonance of AuNPs. This may be caused by polymer swelling in solution or by different extent of PANI doping (at pH 2.7 and pH≈0, correspondingly) which in turn affects the conductivity of PANI but also the PANI chains conformation and thus finally the AuNPs interparticle distance. A decrease of the plasmon absorbance band was observed with increasing potential (from -0.1 V to 0.7 V), and also a small hypsochromic shift and broadening of the absorbance peak. At a higher potential (0.9 V, grey line) the absorbance increases again. A much stronger shift of the absorbance band was observed depending on the oxidation state of PANI when covering gold nanoparticles with a 100 nm thick PANI layer.[35] The absorbance peaks for the two states of PANI were resolved by more than 60 nm and the peak was strongly suppressed in the case of oxidised PANI. This result suggests that the interaction between PANI and AuNPs depends on the specific surroundings of the AuNPs and it becomes more pronounced for the oxidised, high-conducting state of PANI. Thus the shift of the plasmonic band with potential observed in our investigation should be due to the influence of oxidised PANI on the resonance of plasmons. To estimate this effect, the shift of the peak position and of the maximal absorbance was simulated using Mie equation[36] and the data[37] on real and imaginary components of the refractive index of gold. It was found that the observed electrochemically driven hypsochromic shift of the plasmonic absorption peak corresponds to a decrease of the refractive index of PANI by 0.07. A similar observation of electrochemical control of the plasmonic absorption peak was reported earlier for AuNPs deposited between two layers of electropolymerized PANI.[38] It was suggested that the modification of the plasmonic absorbance band relates to changes of the real and imaginary parts of the dielectric constant of PANI. Such changes of the dielectric constants were measured directly by means of surface plasmon resonance spectroscopy in Kretschmann configuration.[39]

Results and Discussion

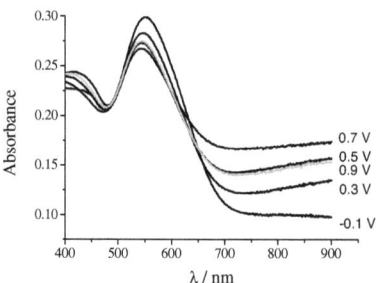

Figure 32. Absorption spectra of four bi – layers of PANI – AuNPs measured at different potentials vs. SCE in 0.5 M sulfuric acid.

Fig. 33 presents cyclic voltammetry curves measured in 0.5 M H_2SO_4 after adsorption of each subsequent PANI or AuNPs layer. The potentiodynamic curves demonstrate the usual oxidation behaviour of polyaniline, i.e. the leucoemeraldine to emeraldine (first anodic peak) and emeraldine to pernigraniline (peak starting at 0.6 V) oxidations and the corresponding reverse reduction processes. The appearance of hydrogen reduction currents close to the negative potential scan limit gives an evidence for the electrocatalytic activity of the gold nanopartcles, immobilised in the LbL structure.

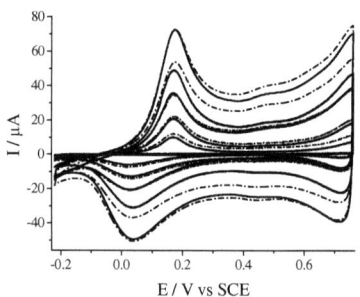

Figure 33. Voltammetric curves measured in 0.5 M H_2SO_4 for the first five bi-layers of PANI – AuNPs after every PANI (solid line) and AuNPs (dotted line) deposition step.

It is interesting to note the increase in the electrochemical activity of PANI after each AuNPs adsorption step (dotted lines). This effect could be attributed to additional chemical doping of the polymer by the adsorbed AuNPs. A similar conclusion was made by Zou et al.[40] for PANI – mercaptoethane sulfonate-stabilized AuNPs and AgNPs, based on comparative UV-vis – near-IR investigations of the LbL assembled PANI structures. The authors suggest that AuNPs and AgNPs provide more effective doping of PANI and improve its conductivity. Fig. 34 shows the dependence of the oxidation charge of the multilayered composite structure, calculated from the cyclic voltammograms, on the number of deposited bi-layers. Dialyzed and non-dialyzed AuNPs solutions were used for the AuNPs adsorption step. The comparison shows a somewhat stronger increase in Q_{ox} in the case of non-dialyzed AuNPs solution, probably due to the higher amount of citrate adsorbed on the Au particles surface. This should result in enhancement of the negative surface charge of the AuNPs, which makes the interaction between the positively charged PANI chains and AuNPs more efficient.

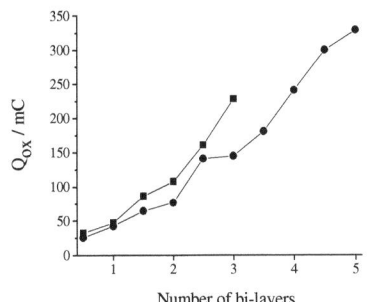

Figure 34. Dependences of Q_{ox} on the number (n) of deposite bi-layers of PANI – AuNPs for dialyzed (squares) and non-dialyzed (circles) gold nanoparticles. The data obtained after deposition of AuNPs correspond to whole numbers of bilayer, the points in between correspond to the data obtained after deposition of PANI

Further voltammetric experiments were carried out in buffer solutions with different acidity (pH from 1 to 7) (Fig. 35). The measurements show well pronounced electrochemical activity of the LbL deposited layers in the pH range 1 – 6. With increasing pH the two intrinsic PANI redox couples gradually shift together and finally, at pH 5 and 6, single broad oxidation and reduction peaks are observed.

Results and Discussion

Between pH 6 and 7 the LbL assembled PANI structure looses almost completely its electrochemical activity, probably due to the transformation from emeraldine-salt to emeraldine-base. Similar observations were made for LbL assembled PANI - AuNPs stabilized with mercaptosuccinic acid [22].

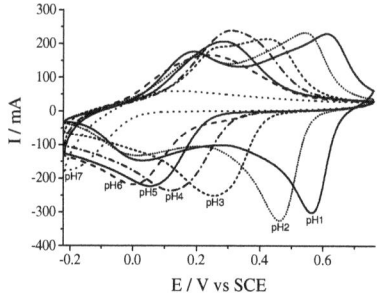

Figure 35. Voltammetric curves of PANI – AuNPs nanocomposite consisting of ten bi-layers measured in buffer solutions at different pH.

The "in-situ" conductance of a PANI – AuNPs LbL structure consisting of 10 bi-layers was investigated at different fixed potentials in buffer solutions at pH 2 - 7 (Fig. 36). The PANI-AuNPs nanocomposite gradually looses its conductivity with increasing pH. Also the potential window corresponding to the high conductivity state of PANI becomes narrower. This is in accordance with the shift of the emeraldine to the pernigraniline oxidation peaks, observed in the voltammetric experiments.

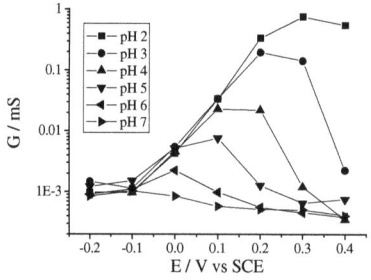

Figure 36. Potential dependence of conductance of PANI – AuNPs nanocomposite consiting of six bi-layers measured in buffer solutions at different pH.

Results and Discussion

The gold surface has a high affinity to mercury vapor and to some sulfuric compounds (thiols, disulfides). Therefore one can expect that binding these compounds by the gold nanoparticles incorporated in the PANI – AuNPs nanocomposites results in a modification of the physical properties of the composite material. The conductometric response of the nanocomposite multilayer structures to vapors of mercury, octanthiol and dimethyldisulfide were measured. Figure 37 A shows the response of a PANI – AuNPs LbL-coated interdigitated electrode on injection of 10 mL of the headspace of octanthiol and dimethyldisulfide performed in a 2 mL flow cell. The observed drop in resistance was specific for the PANI – AuNPs nanocomposite multilayer and was not observed for the electrodes coated by PANI / PSS multilayer films. Therefore the binding of these compounds to the gold nanoparticles is responsible for the resistance decrease. A similar behavior was found on exposure of the nanocomposite to mercury vapor (Fig. 37 B). Again, this effect was not observed for PANI / PSS films.

Both examples show that binding of analytes by gold nanoparticles incorporated in polyaniline can be detected by in-situ resistance measurements. For non-modified gold nanoparticles this effect is not selective but one can expect that a treatment of the surface by deposition of selectively permeable monomolecular layers[30] or by immobilization of receptors on the gold surface can lead to highly selective chemoresistive properties.

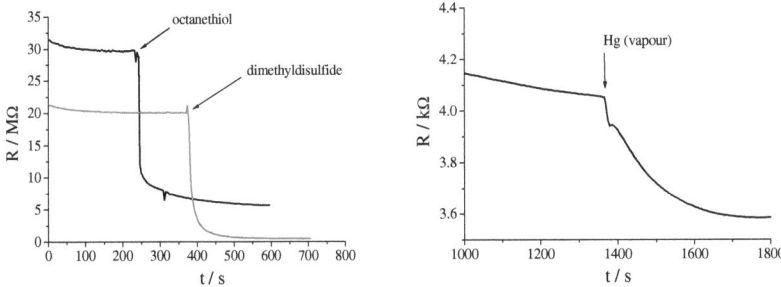

Figure 37. Conductometric response of a nanocomposite consisting of ten bi-layers of PANI – AuNPs on exposure to octanethiol and dimethyldisulfide (a) and mercury (b) vapors measured in two-point configuration.

The possibility to obtain multilayered structures of PANI and AuNPs was demonstrated. The proposed LbL approach results in the formation of PANI shells

63

Results and Discussion

around AuNPs that prevents their further agglomeration. The spectroelectrochemical measurements of PANI – AuNPs nanocomposite show that the physical properties of this nanocomposite multilayer structure cannot be described as a superposition of properties of both components but result from the interaction of AuNPs and the polymer.

The high sensitivity of the resistance of the synthesized PANI-AuNPs nanocomposite on exposure to compounds with high affinity to gold gave evidence for their binding to the Au surface.

3.5.2. Polyaniline palladium nanoparticle composite

Palladium nanoparticles (PdNPs) are known to posses a high electrocatalytic activity towards the electrooxidation of hydrazine.[41]-[44] Modified electrodes containing Pd were obtained either by electrodeposition[45]-[49] or by chemical reduction of Pd salts.[50]-[52] Pre-synthesized Pd NPs were only used in few cases, e.g. for incorporation in polypyrrole in the course of electropolymerization[53] and for anchoring on carbon nanotubes.[54] In this chapter the LbL adsorption technique is used for the preparation of Pd NPs – PANI nanocomposite layers and the possibility to involve the synthesized nanocomposite material for electrochemical sensing of hydrazine in neutral media is explored.

3.5.2.1. Results and Discussion

Upon reduction of $Pd(NO_3)_2$ by ascorbate the initial yellow solution became dark brown indicating the formation of PdNPs. According to the dynamic light scattering, the hydrodynamic size of PdNPs obtained after ultrasound treatment of the solution is about 5–12 nm. TEM images (Fig. 38) indicate a size of 4–10 nm. Dynamic light scattering results obtained in non-sonicated solution show larger sizes most probably due to PdNPs aggregation.

Results and Discussion

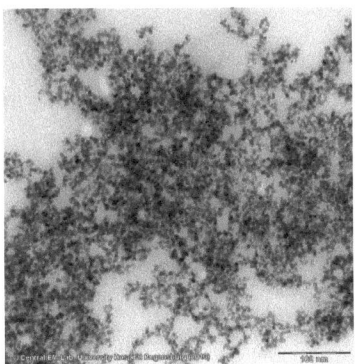

Figure 38. TEM image of the Palladium nanoparticles.

The consecutive adsorption of PANI and PdNPs in the course of the LbL-deposition procedure was followed by cyclic voltammetry carried out in 0.5 M $HClO_4$. The voltammetric response was usually measured after each PdNPs adsorption step, i.e. after each bilayer of PANI and PdNPs (Fig. 39 A). The voltammograms show the growing amounts of both PANI (a continuous increase in the intrinsic PANI oxidation / reduction currents in the potential interval 0.0 to 0.8 V) and PdNPs (a continuous increase in the hydrogen adsorption / desorption currents on Pd at potentials more negative than 0.0 V) after each bilayer adsorption. Independent evidence for the presence of the PdNPs in the LbL deposited composite material was obtained by EDX analysis. It is important to stress that after a step of PdNPs adsorption the consecutive step of PANI adsorption does almost not affect the electrochemical reactivity of the PdNPs. This is demonstrated in Fig. 39 B showing cyclic voltammograms registered after three consecutive adsorption steps: after PANI adsorption for initiating the building of the second bilayer; after PdNPs adsorption completing the building of the second bi-layer and after PANI adsorption initiating the building of the next third bilayer. It can be seen that the adsorption of PdNPs leads to a strong increase in the Pd-based hydrogen adsorption / desorption currents (grey line in the Fig. 2 B). After the next adsorption step of PANI however, only a slight decrease in the hydrogen adsorption / desorption current is observed which means that the adsorbed polymer layer does not affect the reactivity of the PdNPs. Therefore it should be expected that the reactive Pd surface will increase by

65

Results and Discussion

increasing the number of adsorption steps in the nanocomposite preparation procedure.

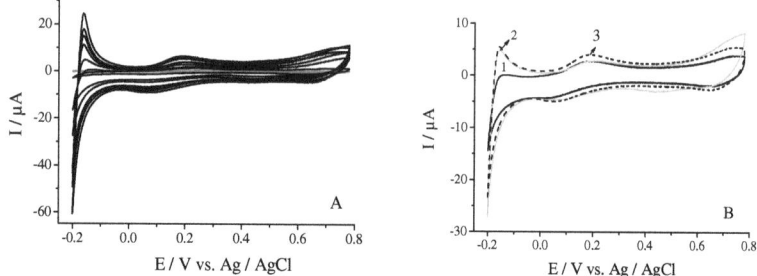

Figure 39. Cyclic voltammograms measured after the deposition of each bilayer in the course of the formation of a 7-bilayers PANI-PdNPs nanocomposite structure on glassy carbon electrode. The inner grey curve shows the electrochemical response of the bare electrode. The signals increased monotonously with the number of bilayers. (A) Cyclic voltammograms registered after three consecutive adsorption steps: after: PANI adsorption for initiating the building of the second bilayer (solid black line); after PdNPs adsorption completing the building of the second bi-layer (solid grey line) and after PANI adsorption initiating the building of the next third bilayer (dashed line). (B) Electrolyte: 0.5 M $HClO_4$. Sweep rate: 100 mV/s.

The SEM imaging of the surface of a 7 bilayer PANI-PdNPs composite (Fig. 40) shows a rough and non-uniform surface morphology of the nanocomposite layer. It could be assumed that the non-uniform surface coverage of the electrode-substrate occurs in the very first adsorption steps. Due to their small size (< 10 nm) no individual Pd nanoparticles could be detected at the magnifications used for the imaging.

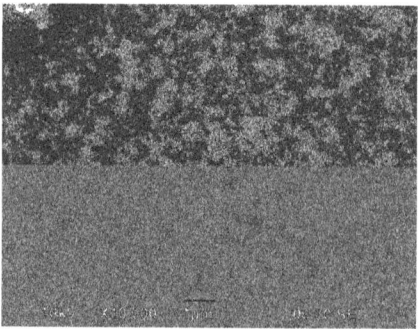

Figure 40. Scanning electron microscopy of a PdNPs-PANI composite formed by LbL adsorption of seven bilayers on glassy carbon electrode.

Further characterization of the PANI-PdNP composite material was carried out by in-situ conductance measurements in a wide range of potentials at two different values of pH. Nanocomposites consisting of 10 bilayers were prepared by using two dilutions of the PANI solutions involved in the polycations adsorption steps. The two sets of data (for 1:10 and 1:100 dilution of the PANI solution) are shown in Fig. 41. The typical potential dependence of the conductance of PANI was observed for the PdNP-PANI composite obtained from the more concentrated solution of PANI (circles in Fig. 41 A and B). As expected for PANI, the conductance diminished with increasing pH and the potential dependence, well pronounced for acidic solutions, was smoothened at higher values of pH. Surprisingly, the PdNP-PANI composite prepared by using the more diluted solution of PANI (squares in Fig. 41 A and B) showed a metal-like behaviour, i.e. almost no potential or pH dependence of its conductance, thus indicating a high contribution of the PdNPs to the composite conductance. It could be assumed that in this case the PdNPs are in close contact with each other and form conducting pathways in the nanocomposite. The metal-like conductance of the PdNPs-PANI nanocomposite, obtained by using the more diluted PANI solution, presents special interest in view of potential application of the nanocomposite material as electrocatalyst in neutral media.

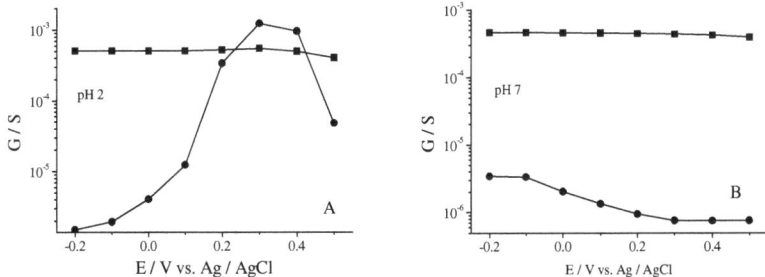

Figure 41. Potential dependence of electrical conductance of Pd NPs-PANI nanocomposite at pH 2 (A) and pH 7 (B). The nanocomposite was formed by LbL adsorption of ten bilayers from concentrated (■) and diluted (●) solutions of PANI.

The cyclic voltammetric behaviour of PdNPs-PANI coated glassy carbon electrodes in neutral solution is shown in Fig. 42 for three nanocomposites consisting of 3, 5 and 7 bilayers. For comparison the response of a Pd-coated glassy carbon electrode is also shown. (Pd was deposited potentiostatically directly on the glassy carbon

Results and Discussion

electrode). Due to the strongly suppressed electroactivity of PANI in this solution the observed voltammograms measured at the PdNPs-PANI nanocomposites reflect mainly the electroactivity of the PdNPs. Only in the case of 3 bilayers the redox activities of both PANI and PdNPs can be clearly seen. The Pd oxide reduction peak appears at 0.1 V, whereas the PANI reduction peak is close to 0 V. With increasing number of adsorption steps, the currents due to Pd-based hydrogen absorption / desorption and Pd oxide formation / reduction increase indicating an increasing content of PdNPs. By monitoring the Pd oxide reduction peak at 0.1 V it becomes evident that the active surface area of the adsorbed PdNPs becomes larger than the surface area of electrodeposited Pd when exceeding 3 PdNP adsorption steps in the formation of the nanocomposite structure.

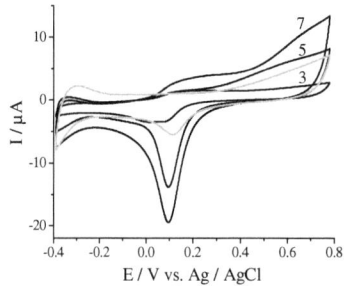

Figure 42. Cyclic voltammograms of PdNPs-PANI nanocomposites consisting of different numbers of bilayers measured in PBS at pH 6.8 (black curves). The number of deposited bilayers is indicated in the figure. The grey curve denotes the voltammetric curve measured at Pd-coated glassy carbon electrode. Sweep rate: 50 mV/s.

The electroactivity of the PdNPs - PANI nanocomposites was further studied with respect to the hydrazine oxidation reaction. A set of cyclic voltammograms registered by using a 7 bilayer nanocomposite at various concentrations of hydrazine in the range 40 µM to 800 µM is presented in Fig. 43 A. The figure shows the stationary voltammograms obtained after several (typically 5) cycles carried out at each concentration of hydrazine in the solution. The effect of the first several voltammetric scans resulting in a gradual decrease in the hydrazine oxidation currents down to a stabilized voltammetric response has been observed in previous studies addressing PdNP-modified electrodes.[45],[48] A gradual decrease within a much larger number

(15 to 30) of scans was observed and commented in terms of stripping of nanoparticles from the electrode surface.[48] For the nanocomposite presented here the stabilization needed only about 5 voltammetric scans (Fig. 43 B) and moreover the almost constant Pd oxide reduction currents did not indicate a loss in the PdNPs immobilized within the PANI material. The large difference between the first scan and the subsequent scans is probably a result of oxidation of the hydrazine pre-concentrated at the electrode surface before the first scan leading to a much higher current. Adsorbed hydrogen is suggested to be an intermediate product of hydrazine electrooxidation.[55] The additional hydrogen species formed during hydrazine oxidation in the former scan should interfere with the hydrazine oxidation and contribute to the oxidation currents in the next scan. Such behaviour was described for bulk palladium.[55] The adsorbed hydrogen on the Pd-nanoparticles probably leads to a shift of the peak potential of later scans in comparison to the first scan in the cyclic voltammogram as a result from hydrogen desorption currents.

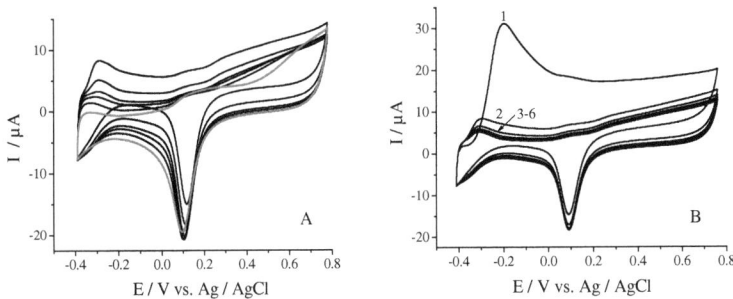

Figure 43. Stationary voltammetric curves measured at a PdNP-PANI nanocomposite at increasing concentrations (40µM, 120µM, 200µM, 400µM, 800µM) of hydrazine (A) and development of the voltammetric curves during the first six voltammetric cycles (indicated by numbers) measured in the presence of 400 µM hydrazine. (B) The nanocomposite consisted of 7 bilayers. Electrolyte: PBS, pH 6.8. Sweep rate: 50 mV/s.

In order to resolve in a better way the hydrazine-related oxidation peaks a subtraction procedure was further used. The reference voltammogram measured in the absence of hydrazine was subtracted from the voltammograms obtained in the presence of hydrazine. The resulting curves (Fig. 44 A) obtained for 7 bilayers of the PANI-PdNP nanocomposite show the emergence of an oxidation peak centered at about -0.2 V.

Results and Discussion

Similar experiments were carried out by using PANI-PdNP nanocomposites of 3 and 5 bilayers. Fig. 44 B shows a comparison of the subtracted voltammetric curves for the three types of composites at constant concentration of hydrazine (200 µM). The hydrazine oxidation peaks shift to lower potential with increasing number of bilayers of the nanocomposites. For the 3 bilayer nanocomposite the current maximum is at about -0.12 V whereas it shifts to -0.28 V for 7 bilayers. In the latter case the value of the peak potential is more negative than the values typically reported for hydrazine oxidation using Pd – based materials.[41]-[44]

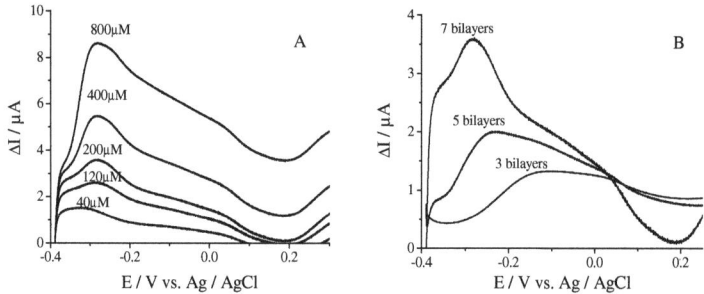

Figure 44. Anodic parts of the voltammetric curves after base line subtraction measured: at 7-bilayered Pd NPs-PANI nanocomposite structure for different hydrazine concentrations (A) and at nanocomposites composed of different numbers of bilayers at a constant (200 µm) hydrazine concentration (B).

In a further experiment the 7 bilayers Pd NPs - PANI nanocomposite was tested for its amperometric response at constant potential in the course of stepwise addition of hydrazine (Fig. 45). In order to obtain a minimum background current the measurements were carried out at 0.2 V. As shown in Fig. 45 A, the addition of hydrazine to a stirred phosphate buffer (pH 6.8) leads to a fast increase of the current. A linear dependence of the current response with increasing concentration was established in the 10 µM to 300 µM range (Fig. 45 B).The sensitivity of the Pd NPs - PANI nanocomposite layer was found to be 0.5 µA / µmol cm^{-2} and the detection limit, calculated from three times signal to noise ratio, was estimated to be 62 nM. This value is below the detection limits communicated in most works addressing hydrazine detection by means of Pd NPs modified electrodes.[41]-[44] A significant increase in the noise was observed at hydrazine concentrations higher than 300 µM.

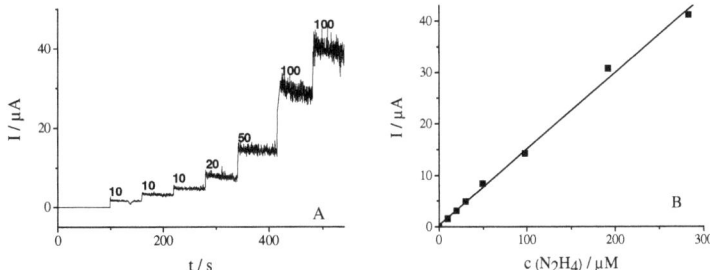

Figure 45. Current response measured at +0.2 V under continuous stirring upon a stepwise increase of the hydrazine concentration (the numbers indicate the concentration increase in µM) (A) and concentration dependence of the current (B). The Pd NPs- PANI nanocomposite is composed of 7 bilayers. Electrolyte: Phosphate buffer 0.1 M, pH 6.8.

Layer by Layer deposition provides the possibility for a simple modification of electrodes with a new electrocatalytically active nanocomposite based on PANI and PdNPs. The immobilization of the PdNPs within the adsorbed PANI layers does not affect the electrochemical reactivity of the PdNPs. Thus the proposed approach offers the possibility to obtain thin nanocomposite layers with reactive three-dimensionally distributed metallic NPs. The concentration of PANI in the solution used for layer by Layer deposition is found to influence significantly the conductance properties of the PdNPs – PANI nanocomposite layers.

The PdNPs – PANI nanocomposite is suitable for electrocatalytic oxidation of hydrazine at a very low potential (-0.28 V). Amperometric measurements at constant potential provided the possibility to obtain a linear response in the 10 to 300 µM range with a sensitivity of 0.5 µA / µmol cm^{-2} and a very low detection limit estimated to be 0.06 µM.

The results obtained together with the investigations on LbL deposition of AuNPs – PANI demonstrate that the suggested approach provides the possibility to obtain a high content of electrochemically active metallic NPs within a non-interfering polymer matrix.

Results and Discussion

3.5.3. Experimental

PANI synthesis

PANI was synthesized by chemical oxidative polymerization of 0.1 M aniline in 0.5 M by addition of 0.1 M ammonium persulfate and subsequent stirring for 6 h at 4°C. The polyaniline, synthesized as emeraldine salt, was transformed to emeraldine base by treatment with ammonia. The obtained product was washed, dried and dissolved in dimethylacetamide (20 mg dried emeraldine base per ml) under continuous stirring and subsequent sonication.[2] After filtering with Millipore filter (0.5 μm), the solution was diluted 10 or 100 times using diluted HCl (pH 3.1). Finally, the pH of the PANI containing solution was adjusted to pH 2.6.

Gold nanoparticle synthesis

The gold nanoparticle solution was prepared by fast injection of 500 μl of 2 % (w/w) sodium citrate in water to 10 ml of a boiling solution of 1 mM $HAuCl_4$ in water under rapid stirring. Stirring was continued until the typical red coloration of the mixture appeared and was followed by cooling down to room temperature. The excess of citrate was removed by diafiltration using a Masterflex tangential flow filtration system (www.pall.com) with a membrane (Omega 10K Membrane, Minimate TFF Capsule, www.pall.com), keeping the volume of the AuNP – solution constant by addition of Millipore water.

Palladium nanoparticle synthesis

10 ml of a 1 mM solution of $Pd(NO_3)_2$ in water was heated until boiling. Then 1 ml of a 0.1 M solution of sodium ascorbate in water was quickly added. The solution was further boiled for about 3 min until the solution changed its colour to grey-brown and then allowed to cool down at room temperature. The particles were characterized by dynamic light scattering using a Zetasizer Nano (Malvern Instruments Ltd.) by transmission electron microscopy (TEM) on a LEO 912 TEM and by UV–Vis spectroscopy.

Layer by layer deposition

Two different PANI solutions were used for adsorption – 1/10 and 1/100 v/v diluted with respect to the original PANI dimethylacetamide organic solution. For LbL deposition the AuNP solution was used without dilution and the Pd NPs solution was 1:1 diluted by water. The deposition started by dipping the substrates first in the PANI solution and then in the nanoparticle solution for 20 min for each step. For the next

Results and Discussion

layers the deposition time was reduced to 10 min. After each adsorption step the electrode was washed in HCl aqueous solution (pH 2.6).The substrates used for the formation of the Nanoparticle–PANI nanocomposite layers were glassy carbon (GC) or indium-tin-oxide (ITO) electrodes for the electrochemical measurements, glass or ITO slides for the optical measurements and interdigitated gold electrodes for the in situ conductance measurements.

An adsorption of PANI onto gold surface was also confirmed by surface plasmon resonance technique, using SPR spectrometer Biosuplar-410 (www.biosuplar.com). The adsorption of the first PANI layer to ITO and glassy carbon was confirmed by cyclic voltammetry in 0.5 M H_2SO_4 (Fig. 4). In these cases a usual physical adsorption of polymers to many surfaces can be enhanced by interaction of polymer charges with induced charges in the conducting phase. The subsequent adsorption of PANI and metal nanoparticles is based on electrostatic attraction between oppositely charged species.

Characterisation of the PANI metal nanoparticle nanocomposites

All electrochemical measurements were carried out with Autolab PGSTAT-12 (www.ecochemie.nl) in a three electrode configuration. Saturated calomel reference electrodes were used and platinum plates served as counter electrodes. Investigations of the pH dependence of the electrochemical activity and conductance of the multilayer PANI-based structures were carried out in buffer solutions with the following compositions: i) 0.1 M NaCl + 10 mM NaH_2PO_4 / H_3PO_4 for pH range from 2 to 3 ii) 0.1 M NaCl + 10 mM CH_3COOH / CH_3COONa for pH range from 4 to 5; iii) 0.1 M NaCl + 10 mM Na_2HPO_4 / NaH_2PO_4 for pH range from 6 to 7. Deionized water additionally purified by Millipore Milli Q system was used for the preparation of all solutions.

TEM imaging of the PANI AuNP compsite was carried out using JEOL JEM 100 B transmission electron microscope. The specimens for TEM were prepared by LbL self-assembly directly on Formvar®-coated Au mesh for TEM. UV-vis absorption spectroscopy and spectroelectrochemical measurements were performed using a Cary 50Bio spectrophotometer from Varian in a 1 cm cell. A two-electrode configuration with Ag/AgCl electrode was used for spectroelectrochemical measurements.

Resistance measurements were carried out as desribed in chapter 2. In the investigation of the conductometric response of the PANI AuNP composite towards

Results and Discussion

octanthiol and dimethyldisulfide the composite was deprotonated in a pH 7 buffer and dried before measurements. To measure the response towards octanthiol and dimethyldisulfide, 10 mL of the headspace of these compounds was passed through a 2 mL flow cell in which the sensor was placed. For investigations of the response towards mercury, the nanocomposite was protonated by pretreatment at pH 2.6 and dried. Then a small droplet of mercury was placed in the sealed measurement chamber with a 1µL pipette.

The electrooxidation of hydrazine on Pd NPs–PANI nanocomposite layers was investigated by of cyclic voltammetry and chronoamperometry in a 100 mM phosphate buffer solution (pH 6.8) by adding different amounts of a 5 mM stock solution of hydrazine in phosphate buffer. Chronoamperometric measurements were carried out at 0.2 V vs. Ag / AgCl under constant stirring.

3.5.4. References

[1] G. Decher, J. D. Hong, J. Schmitt, Thin Solid Films **1992**, 210-211, 831-835.
[2] M. Ferreira, J. Cheung, M. Rubner, Thin Solid Films **1994**, 244, 806-809.
[3] J. Cheng, A. Fou, M. Rubner, Thin Solid Films **1994**, 244, 985-989.
[4] L. H. C. Mattoso, V. Zucolotto, L. G. Patterno, R. van Griethuijsen, M. Ferreira, S. P. Campana, O. N. Oliveira, Synth. Met. **1995**, 71, 2037-2038.
[5] M. Ferreira, M. F. Rubner, Macromolecules **1995**, 28, 7107-7114.
[6] N. Sarkar, M. Ram, A. Sarkar, R. Narizzano, S. Paddeu, C. Nicolini, Nanotechnol. **2000**, 11, 30.
[7] T. Jung, Electroanal. **2003**, 15, 1453-1459.
[8] N. Kovtyukhova, P. J. Ollivier, S. Chizhik, A. Dubravin, E. Buzaneva, A. Gorchinskiy, A. Marchenko, N. Smirnova, Thin Solid Films **1999**, 337, 166-170.
[9] N. I. Kovtyukhova, A. D. Gorchinskiy, C. Waraksa, Mater. Sci. Eng. B **2000**, 69-70, 424-430.
[10] N. I. Kovtyukhova, B. R. Martin, J. K. N. Mbindyo, T. E. Mallouk, M. Cabassi, T. S. Mayer, Mater. Sci. Eng. C **2002**, 19, 255-262.
[11] F. Huguenin, M. Ferreira, V. Zucolotto, F. C. Nart, R. M. Torresi, O. N. Oliveira, Chem. Mater. **2004**, 16, 2293-2299.
[12] W. B. Stockton, M. F. Rubner, Macromolecules **1997**, 30, 2717-2725.

[13] Y. Wang, C. Guo, Y. Chen, C. Hu, W. Yu, J. Colloid. Interface Sci. **2003**, 264, 176-183.

[14] K. Y. K. Man, H. L. Wong, W. K. Chan, A. B. Djurišić, E. Beach, S. Rozeveld, Langmuir **2006**, 22, 3368-3375.

[15] V. Zucolotto, M. Ferreira, M. R. Cordeiro, C. J. Constantino, W. C. Moreira, J. Oliveira, Sens. Actuators B **2006**, 113, 809-815.

[16] R. Dronov, D. G. Kurth, H. Möhwald, F. W. Scheller, F. Lisdat, Electrochim. Acta **2007**, 53, 1107-1113.

[17] R. Dronov, D. G. Kurth, H. Möhwald, F. W. Scheller, J. Friedmann, D. Pum, U. B. Sleytr, F. Lisdat, Langmuir **2008**, 24, 8779-8784.

[18] R. Spricigo, R. Dronov, K. V. Rajagopalan, F. Lisdat, S. Leimkuhler, F. W. Scheller, U. Wollenberger, Soft Matter **2008**, 4, 972-978.

[19] X. Yu, G. A. Sotzing, F. Papadimitrakopoulos, J. F. Rusling, Anal. Chem. **2003**, 75, 4565-4571.

[20] K. Loh, J. Lynch, N. Kotov, Smart Structures and Systems **2008**, 4, 531-548.

[21] S. Tian, J. Liu, T. Zhu, W. Knoll, Chem. Commun. **2003**, 2738-2739.

[22] S. Tian, J. Liu, T. Zhu, W. Knoll, Chem. Mater. **2004**, 16, 4103-4108.

[23] X. Luo, A. Morrin, A. Killard, M. Smyth, Electroanal. **2006**, 18, 319-326.

[24] J. M. Kinyanjui, J. Hanks, D. W. Hatchett, A. Smith, M. Josowicz, J. Electrochem. Soc. **2004**, 151, D113-D120.

[25] M. Sheffer, D. Mandler, Electrochim. Acta **2009**, 54, 2951-2956.

[26] J. A. Smith, M. Josowicz, J. Janata, J. Electrochem. Soc. **2003**, 150, E384-E388.

[27] D. W. Hatchett, M. Josowicz, Chem. Rev. **2008**, 108, 746-769.

[28] J. Turkevich, P. C. Stevenson, J. Hillier, Discuss. Faraday Soc. **1951**, 11, 55-75.

[29] J. J. McNerney, P. R. Buseck, R. C. Hanson, Science **1972**, 178, 611-612.

[30] V. Mirsky, M. Vasjari, I. Novotny, V. Rehacek, V. Tvarozek, O. Wolfbeis, Nanotechnol. **2002**, 13, 175.

[31] M. Daniel, D. Astruc, Chem. Rev. **2004**, 104, 293-346.

[32] A. C. Templeton, J. J. Pietron, R. W. Murray, P. Mulvaney, J. Phys. Chem. B **2000**, 104, 564-570.

[33] J. Schmitt, P. Mächtle, D. Eck, H. Möhwald, C. A. Helm, Langmuir **1999**, 15, 3256-3266.

[34] C. Jiang, S. Markutsya, V. V. Tsukruk, Langmuir **2004**, 20, 882-890.

[35] Y. R. Leroux, J. C. Lacroix, K. I. Chane-Ching, C. Fave, N. Félidj, G. Lévi, J. Aubard, J. R. Krenn, A. Hohenau, J. Am. Chem. Soc. **2005**, 127, 16022-16023.
[36] G. Mie, Annalen der Physik **1908**, 330, 377-445.
[37] L. Schulz, F. Tangherlini, J. Opt. Soc. Am. **1954**, 44, 362-367.
[38] Y. Leroux, E. Eang, C. Fave, G. Trippe, J. C. Lacroix, Electrochem. Comm. **2007**, 9, 1258-1262.
[39] A. Baba, S. Tian, F. Stefani, C. Xia, Z. Wang, R. C. Advincula, D. Johannsmann, W. Knoll, J. Electroanal. Chem. **2004**, 562, 95-103.
[40] X. Zou, H. Bao, H. Guo, L. Zhang, L. Qi, J. Jiang, L. Niu, S. Dong, J. Colloid Interface Sci. **2006**, 295, 401-408.
[41] J. Liu, W. Zhou, T. You, F. Li, E. Wang, S. Dong, Anal. Chem. **1996**, 68, 3350-3353.
[42] T. Li, E. Wang, Electroanal. **1997**, 9, 1205-1208.
[43] C. Batchelor-McAuley, C. E. Banks, A. O. Simm, T. G. J. Jones, R. G. Compton, Analyst **2006**, 131, 106-110.
[44] R. Baron, B. Sljukic, C. Salter, A. Crossley, R. Compton, Electroanal. **2007**, 19, 1062-1068.
[45] F. Li, B. Zhang, S. Dong, E. Wang, Electrochim. Acta **1997**, 42, 2563-2568.
[46] D. Guo, H. Li, Electrochem. Comm. **2004**, 6, 999-1003.
[47] D. Guo, H. Li, J. Colloid Interface Sci. **2005**, 286, 274-279.
[48] X. Ji, C. Banks, A. Holloway, K. Jurkschat, C. Thorogood, G. Wildgoose, R. Compton, Electroanal. **2006**, 18, 2481-2485.
[49] N. Maleki, A. Safavi, E. Farjami, F. Tajabadi, Anal. Chim. Acta **2008**, 611, 151-155.
[50] B. Dong, B. He, J. Huang, G. Gao, Z. Yang, H. Li, J. Power Sources **2008**, 175, 266-271.
[51] Y. Shen, Q. Xu, H. Gao, N. Zhu, Electrochem. Commun. **2009**, 11, 1329-1332.
[52] C. Shao, N. Lu, Z. Deng, J. Electroanal. Chem. **2009**, 629, 15-22.
[53] C. R. Rao, D. Trivedi, Catal. Commun. **2006**, 7, 662-668.
[54] L. Chen, G. Hu, G. Zou, S. Shao, X. Wang, Electrochem. Commun. **2009**, 11, 504-507.
[55] N. V. Korovin, B. N. Yanchuk, Electrochim. Acta **1970**, 15, 569-580.

3.6. PEDOT / PSS palladium nanoparticle composite

Conducting polymer nanoparticle composites can be prepared by electrochemical deposition of metallic nanoparticles on a surface or in a conducting polymer matrix,[1] by electroless reduction of metal salt precursors by the conducting polymer,[1],[2] by reduction of metal salt precursors by monomers resulting in polymerisation of the monomers,[3] by electropolymerisation of conducting polymers in presence of nanoparticles in solution[4] and by electrostatically driven layer by layer deposition of metal nanoparticles and conducting polymers (chapter 3.5).[5],[6] However there are only few works dealing with the preparation of dispersions from conducting polymers and pre-synthesized nanoparticels.[7] Such dispersions have the advantage that they can be easily transferred to almost any surface by ink-jet printing, spin- or spray-coating.[8]

In this chapter the characterisation of a nanocomposite consisting of PEDOT-PSS and palladium nanoparticels and its application for conductometric detection of hydrazine and NADH is presented. Furthermore, a new approach to calibrate such sensors is introduced.

3.6.1. Results and Discussion

TEM images of the PEDOT-PSS / PdNP composite obtained by adsorption of the composite on a carbon coated copper grid, show 5 to 10 nm big palladium particles (Fig. 46). The characterisation of the nanoparticles itself including TEM and light scattering data is given in the previous chapter (chapter 3.5.2.). It seems that the particles are stacked together in a loose random structure (Fig. 46). It is known that PEDOT-PSS consists of 30 - 200 nm particles,[9],[10] but they scatter the electrons much smaller than the palladium nanoparticles and therefore are not observed in the TEM image.

Results and Discussion

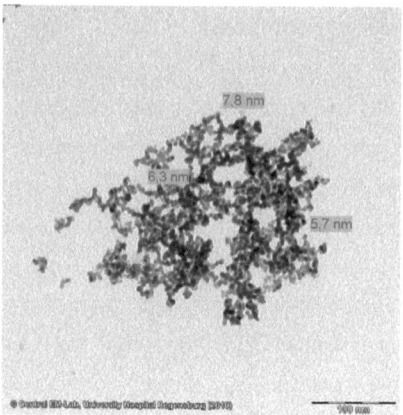

Figure 46. TEM images of palladium particles in the PEDOT-PSS matrix.

Optical microscopy of a film of the composite on an interdigitated electrode shows homogeneously distributed small black particles with the size of around 1 μm along with some bigger aggregates within a thin film of PEDOT-PSS. It has to be noted that the film covers the whole surface (Fig. 47). Furthermore no connections of the microelectrodes by big agglomerates of palladium were found in all experiments.

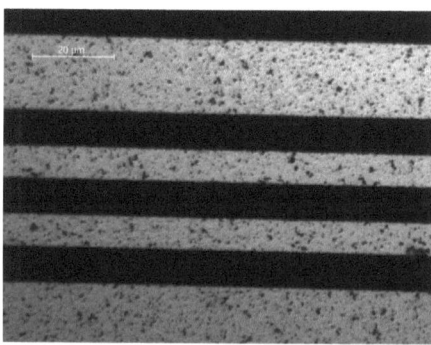

Figure 47. Picture of a thin film of the composite on an interdigitated electrode, acquired by optical microscopy

Cyclic voltammograms were recorded to evaluate the presence, electroactivity and stability of palladium particles in the layers. The voltammogramm shows the typical features of palladium; e.g. the currents corresponding to the hydrogen adsorption (<-0.2 V) and desorption (-0.05 V) and oxide formation (> 0.7 V) and reduction (0.5 V) (Fig. 48 A). The voltammogram is very similar to that of a PEDOT-PdNP composite obtained by electroless deposition of palladium in previously reduced PEDOT films.[2] For comparison, the voltammogram for pure PEDOT-PSS on the glassy carbon electrode is shown in Fig. 48 B. The currents of the PEDOT oxidation are much smaller than the currents observed for hydrogen adsorption / desorption on the palladium nanoparticles. At the starting voltage of -0.35 V PEDOT is already partially oxidized, therefore only partial oxidation and reduction of PEDOT takes place during this voltammetric experiments.

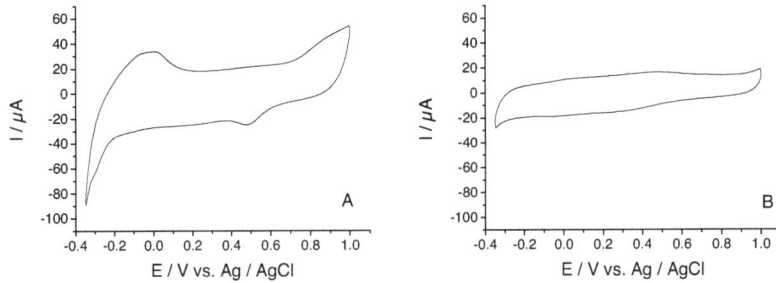

Figure 48. Voltammogram of the nanocomposite of PdNP with PEDOT-PSS (A) and voltammogram of PEDOT-PSS (B). Electrolyte: 0.5 M H_2SO_4. Sweep rate: 0.1 V/s.

The catalytically active palladium nanoparticles are electrically connected by PEDOT-PSS. This is proved by in-situ conductance measurements. The potential dependent conductance of the nanocomposite is very similar to that of PEDOT-PSS without palladium nanoparticels and seems therefore not influenced by the presence of the PdNPs (Fig. 49). This is in contrast to the PANI-PdNP composite obtained by layer by layer deposition from diluted PANI solutions in chapter 3.5.2,[11] where PdNPs are assumed to build conducting pathways in the PANI matrix.

The potential dependent conductance of PEDOT can be described with the simple model described in chapter 3.2. However in contrast to polythiophene only two states a conducting state A and a non-conducting state B are proposed. This results in the following equation for the potential dependence of the conductance.

Results and Discussion

$$G = \frac{g'_A + (g'_B \cdot 10^{\frac{(E_{0\_PEDOT} - E)}{a}})}{1 + 10^{\frac{(E_{0\_PEDOT} - E)}{a}}} \quad (1)$$

where $a = 2.3 \cdot \frac{RT}{n_{AB}F}$.

As one can see in Fig. 49, equation 1 describes the potential dependent conductance of the film quite well, especially in the conducting region. It has to be stated that a logarithmic plot of the conductance versus the potential would give a better view of the potential dependence of the composite as with the linear plot the small changes in the highly oxidized state are overestimated,[12] however in this work the high conducting region is of more interest as even with high concentrations of hydrazine we do not reach potentials less than -0.25 V.

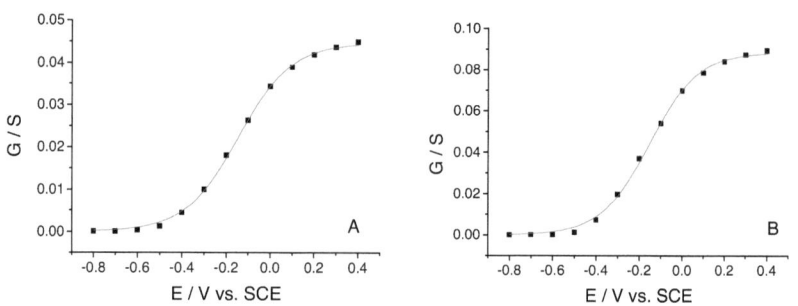

Figure 49. Potential influence on the conductance of the nanocomposite from PdNP with PEDOT-PSS (A) and on PEDOT-PSS without palladium nanoparticles (B). Electrolyte: 100 mM phosphate, pH 7.

The electroactivity of the nanocomposite from PdNP and PEDOT-PSS towards hydrazine oxidation was studied by cyclic voltammetry in a pH 7 buffer. A set of cyclic voltammograms registered by using a glassy carbon electrode coated by the composite at various concentrations of hydrazine in the range 40 µM to 800 µM is presented in Fig. 50. The figure shows the stationary voltammograms obtained after 3 cycles carried out at each concentration of hydrazine in the solution. The oxidation of hydrazine seems to start at about -0.3 V and has a maximum at about 0.15 V.

Results and Discussion

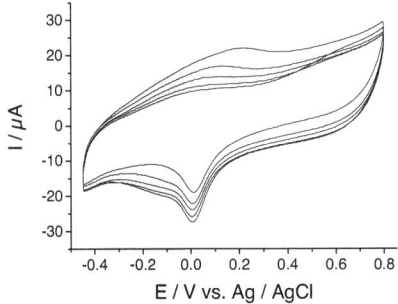

Figure 50. Increased concentrations of hydrazine lead to increase of the current on voltammograms of the nanocomposite of PdNP and PEDOT-PSS deposited on a glassy carbon electrode. The hydrazine concentrations are: 0, 20, 100, 200, 400 µmol/L. Electrolyte: 100 mM phosphate, pH 7. Sweep rate: 0.1 V/s.

As expected, an exposure of the nanocomposite of PdNP with PEDOT-PSS to hydrazine or NADH leads to the reduction of PEDOT accompanied by the decrease in the conductance (Fig 51). It was possible to detect concentrations as low as 0.5 µM for hydrazine and 10µM for NADH. Lower concentrations did not lead to any signal change. This was observed at various measurements with different electrodes coated by the composite, however the response varied from electrode to electrode due to the simple and crude sensor fabrication. It has to be noted that for NADH a thinner layer of the composite was used resulting in a less conducting layer. In comparison to hydrazine the reduction of the composite by NADH was rather slow and the conductance changes where lower. One can assume that hydrazine can diffuse into the composite whereas NADH oxidation probably takes place only on the surface. Probably the PdNPs immobilized in the polymer matrix serve as reaction sites for hydrazine/NADH oxidation and the released electrons reduce PEDOT. This assumption is supported by the fact that the pure PEDOT PSS covered electrode showed only weak changes upon exposure to hydrazine (14 % change at 10mM in comparison to 45 %; see inset Fig. 51 A). It has to be noted that the composite does not only respond to these two analytes, but most probably also to other reducing agents like ascorbic acid etc.

81

Results and Discussion

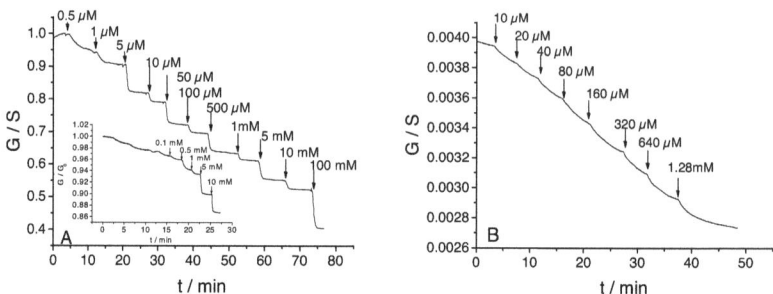

Figure 51. Influence of hydrazine (A) and NADH (B) on the conductance of the nanocomposite from PdNP with PEDOT-PSS. Electrolyte: 100 mM phosphate, pH 7.

Alternatively, the analytical information can be obtained from potentiometric measurements. The open circuit potential of the composite decreased by 39 mV per decade with increased hydrazine concentration (Fig. 52). This value can be used to analyze the stoichiometry of the reaction, however here it is considered as an empiric parameter which can be used for quantitative analysis of the hydrazine concentration. The equation for the potential of the nanocomposite coated electrode is:

$$E = E_0 - 0.039 \cdot \log(c(N_2H_4)) \quad (2)$$

where E_0 = - 0.26 V.

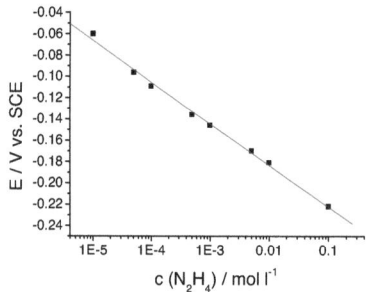

Figure 52. Open circuit potential of the interdigitated gold electrode coated by the nanocomposite of PdNP and PEDOT-PSS as a function of hydrazine concentration.

Taking into account this dependence and the dependence of the conductance of the composite on potential it is clear why the response is not linear with increasing concentration, but almost linear in the logarithmic concentration scale. By assigning the open circuit potential values to the conductance values for each hydrazine concentration and plotting these data together with the potential dependent conductance of the nanocomposite of PdNP and PEDOT-PSS, one can see that the data match each other perfectly (Fig. 53). Thus chemical and electrical control of the redox potential of the film yield quantitatively the same result.

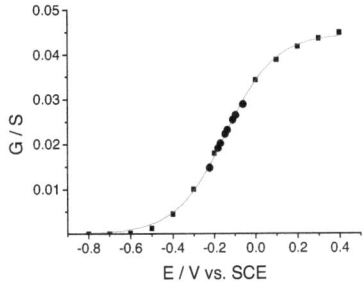

Figure 53. Comparison of the dependencies of the conductance of nanocomposite from PdNP and PEDOT-PSS on the applied electrode potential (squares) and on the open circuit potential defined by hydrazine additions (circles). The line presents the theoretical curve according to the eq 1.

Combining the equation used for the description of the potential dependent conductance of the composite (eq 1) with the equation used for description of the electrode potential in dependence of the hydrazine concentration (eq 2), one gets an equation which describes the dependence of the conductance on the concentration of hydrazine in solution.

$$G = \frac{g_A' + g_B' \cdot 10^{\frac{(E_{0\_PEDOT} - E_{0\_N_2H_4} - 0.039 \cdot \log(c(N_2H_4)))}{a}}}{1 + 10^{\frac{(E_{0\_PEDOT} - E_{0\_N_2H_4} - 0.039 \cdot \log(c(N_2H_4)))}{a}}} \quad (3)$$

Fig. 54 shows that equation 3 describes the concentration dependence of the conductance of the layer. It has to be noted that the all parameters used for this description were determined from the potential dependent conductivity of the

Results and Discussion

composite and the potentiometric response of the composite on increasing hydrazine concentrations. No fitting procedure was used to describe the dependence. The same approach was used for NADH, however in this case the sensitivity of the conductance was 59 mV per decade.

Another approach suggested in literature to calibrate chemiresistors is based on the response time of the sensor after analyte addition.[13]-[15] However this approach requires that the sensor is switched to its initial potential before each concentration determination and is therefore not suitable for continuous monitoring of analyte concentration.

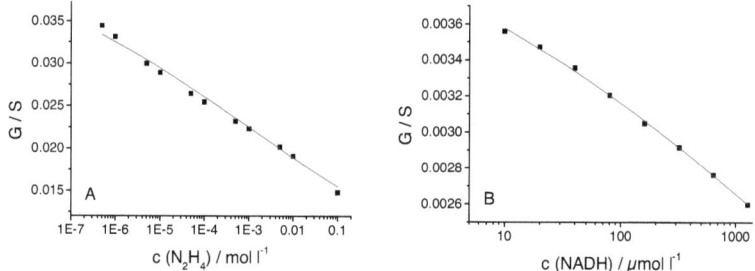

Figure 54. Calibration of the conductometric hydrazine (A) and NADH (B) chemiresistors based on the theoretical model describing red-ox conversions of PEDOT. The squares indicate the measured conductance, whereas the lines represent theoretical dependencies obtained from the eq 5 with corresponding parameters for hydrazine and NADH.

A new chemosensitive nanocomposite material was obtained from palladium nanoparticles and PEDOT-PSS. The composite forms a stable dispersion in aqueous media which can be used for deposition on various surfaces via drop-, spray- or spin-coating. The composite was found to be chemosensitive towards reducing agents like hydrazine or NADH and belongs therefore to the class of broad-selective chemosensitive materials for detection of reducing compounds. Conductometric or potentiometric transducing was used. The chemosensitivity is probably based on electrocatalytical activity of palladium nanoparticles corresponded by the reduction of PEDOT. A simple model describing the redox conversion of PEDOT provided quantitative description of this system and was used for the development of a new calibration approach.

Results and Discussion

3.6.2. Experimental

250 µl of the commercially available PEDOT-PSS dispersion (Sigma Aldrich 1.3 % (w/w), conductive grade) were diluted with 9.75 ml of water. 500 µL of this solution were mixed with the PdNP solution resulting in a stable dispersion of the PdNPs in the PEDOT-PSS solution. 1% of dimethylsulfoxide (DMSO) was added to this solution before deposition onto the electrodes. An addition of DMSO to PEDOT-PSS is known to enhance the conductivity of PEDOT-PSS layers.[16] The chemiresistors were prepared by drop-coating of 0.5 µL of this dispersion on electrodes with a working area of 0.64 mm^2 consisting of four interdigitated gold strips separated by an 8 µm gap (Fig. 1). Then the chemiresistors were dried at about 80°C for 10 min. For the NADH sensor, 0.2 µL instead of 0.5 µL were used. Before coating the electrodes were cleaned subsequently by acetone, ethanol, piranha solution and water.

The samples for cyclic voltammetry were prepared by drop-coating of 6 µl of the composite dispersion on a glassy carbon electrode (3.14 mm^2) and drying at 80°C for 10 min.

Electrochemical measurements were done using a CHI-660A electrochemical analyzer. Either Ag / AgCl electrodes or saturtated calomel electrodes were used as reference electrodes. The type of the electrode used is indicated in the corresponding figures. A platinum wire was used as a counter electrode.

Simultaneous two- and four point measurements were used as described in chapter 2. To avoid effects due to a potential difference between the outer electrodes, only a small voltage difference (10 mV) was applied. Phosphate buffer (pH 7, 100mM) was used in all experiments. Hydrazine hydrate solutions were made by dilution of a 1 M stock solution in buffer. To measure the effect of hydrazine the sensor was placed in a flow cell. Each concentration was pumped through the flow cell till the response of the sensor reached the saturation value. In case of NADH additions of NADH from a 10 mM stock solution do the pH 7 buffer containing the sensor were performed.

3.6.3. References

[1] V. Tsakova, Journal of Solid State Electrochem. **2008**, 12, 1421-1434.
[2] S. Eliseeva, V. Malev, V. Kondratiev, Russ. J. Electrochem. **2009**, 45, 1045-

1051.
[3] J. M. Kinyanjui, N. R. Wijeratne, J. Hanks, D. W. Hatchett, Electrochim. Acta **2006**, 51, 2825-2835.
[4] C. Bose, K. Rajeshwar, J. Electroanal. Chem. **1992**, 333, 235-256.
[5] U. Lange, S. Ivanov, V. Lyutov, V. Tsakova, V. Mirsky, J. Solid State Electrochem. **2010**, 14, 1261-1268.
[6] B. Vercelli, G. Zotti, A. Berlin, J. Phys. Chem. C **2009**, 113, 3525-3529.
[7] R. Pacios, R. Marcilla, C. Pozo-Gonzalo, J. A. Pomposo, H. Grande, J. Aizpurua, D. Mecerreyes, J. Nanosci. Nanotechn. **2007**, 7, 2938-2941.
[8] K. Crowley, A. Morrin, A. Hernandez, E. O'Malley, P. G. Whitten, G. G. Wallace, M. R. Smyth, A. J. Killard, Talanta **2008**, 77, 710-717.
[9] U. Lang, E. Müller, N. Naujoks, J. Dual, Adv. Funct. Mater. **2009**, 19, 1215-1220.
[10] H. Yan, S. Arima, Y. Mori, T. Kagata, H. Sato, H. Okuzaki, Thin Solid Films **2009**, 517, 3299-3303.
[11] S. Ivanov, U. Lange, Vessela Tsakova, V. M. Mirsky, Sens. Actuators B, in press.
[12] J. Heinze, B. A. Frontana-Uribe, S. Ludwigs, Chem. Rev. **2010**, DOI: 10.1021/cr900226k.
[13] P. Bartlett, Y. Astier, Chem. Commun. **2000**, 2000, 105-112.
[14] P. N. Bartlett, P. R. Birkin, Anal. Chem. **1994**, 66, 1552-1559.
[15] P. N. Bartlett, J. H. Wang, E. N. K. Wallace, Chem. Commun. **1996**.
[16] J. Y. Kim, J. H. Jung, D. E. Lee, J. Joo, Synth. Met. **2002**, 126, 311-316.

Results and Discussion

3.7. Graphene based gas sensors

Graphene was shown to be a promising new chemosensitive material for gas sensors.[1],[2] Up to now there are however only few works dealing with its application in gas sensors and all these works can be considered as preliminary studies.[3]-[9] In these chapter graphene obtained by reduction of graphene oxide is characterized and the possibility to use this material in gas sensors is evaluated. Furthermore possibilities to tune the sensitivity and selectivity of graphene by modification with metallic nanoparticles are given.

3.7.1. Graphene characterisation

Characterisation of chemically derived graphene is essential for applications, as its properties depend on many factors e.g. the graphite raw material, the synthesis procedure, the reduction procedure and post treatment procedures.

3.7.1.1. Results and Discussion

Oxidation of graphite by a modified Hummer's method[10],[11] yields a stable yellow-brown solution of graphene oxide. The UV-spectra of the solution is shown in Fig. 55. Upon reduction of grapheneoxide with hydrazine the absorption peak of grapheneoxide at 229 nm redshifted to 265 nm, the shoulder at 300 nm disappeared and the absorption increased in the whole spectral range.[12]

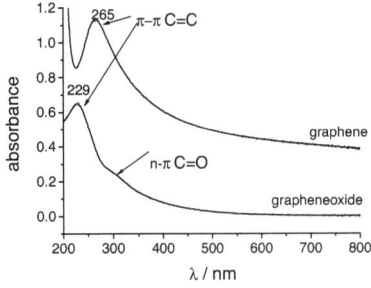

Figure 55. Absorbance spectra of graphene and grapheneoxide in water (c=0.01 g / l).

87

Results and Discussion

Fig. 56 shows the scanning electron microscopy (SEM) pictures of the resulting graphene flakes on an electrode for resistance measurements and a platinum surface at two different magnifications. The deposition was done by dropcoating 1 µl of a 0.01 g / l graphene solution on the electrode on a hotplate at about 80°C. The size of the graphene flakes varies from about 200 nm to > 1µm. From the SEM images one can clearly distinct regions with single flakes and regions of overlapping flakes. An electrode coated in this way was used for measurements of the resistance response upon exposure to NO_2, hydrogen and water vapour.

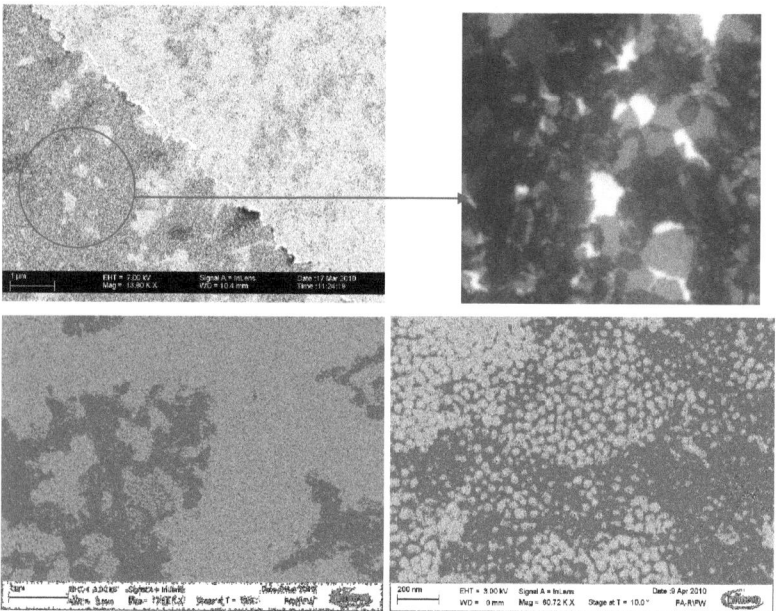

Figure 56. SEM images of graphene flakes dropcoated on an electrode for resistance measurements and on a platinum surface at different magnifications.

The Raman spectra of the graphene flakes obtained with an argon-ion-laser with a wavelength of 488 nm show the D and G band of graphene at 1350 nm and 1611 cm^{-1} (Fig. 57). It has to be noted that the G band consists actually of two bands, the G band at around 1580 cm^{-1} and the D'-band located at 1620 cm^{-1}.[13] Furthermore a number of broad bands occur between 2400 cm^{-1} and 3200 cm^{-1}, which can be assigned as secondary bands of the D, G and D' band and a combination of the G and D band.[14] From the D / G ratio and the position of the G-Band one can estimate

Results and Discussion

the size of the crystalline domains in graphene. According to [13] and [15] the measured ratio of 1.05 and the position of the G-Band at about ~1600 cm^{-1} would yield a size of about 6 nm for the crystalline domains. In comparison to the grapheneoxide Raman-spectra (Fig. 57) the G-band shifted to higher wavenumbers and the D to G ration increased. Furthermore a slight decrease in the D+G to 2D ratio was found. Several groups correlated the the D to G, the D+G to 2 D or the 2D to G band ratio with the conductivity in reduced graphene oxide. For the D to G ratio it was found that it first increases with increasing conductivity, but decreases again at high conductivites, whereas the D+G to 2D ratio decreased and the 2D to G ratio increased with increasing conductivity.[16]-[18]

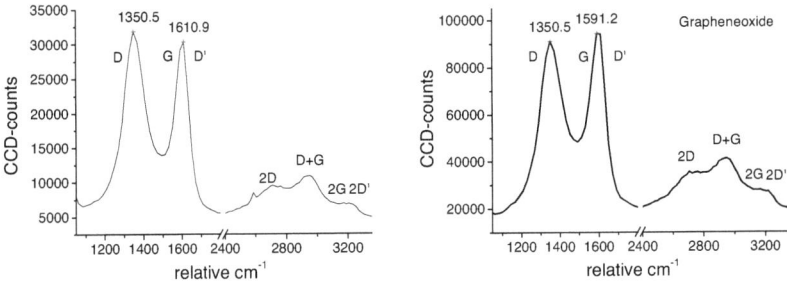

Figure 57. Raman spectra of graphene and grapheneoxide. Laser Ar-ion: λ=488 nm.

It is known from literature that grapheneoxide can be converted to graphene by heat treatment under Argon or Ar / H$_2$ atmosphere or under vacuum.[16],[19],[20] Furthermore heating chemically reduced graphene yields a material with improved conductivity.[11],[21] To investigate this effect and to evaluate the temperature stability of graphene and grapheneoxide, both substances were investigated by thermogravimetric analysis (TGA). Additionally one interdigitated electrode was coated by graphene (2 µL / 0.01 mg / mL) and subjected to a thermal treatment at 600°C under vacuum. The resistance of this electrode before the thermal treatment was 400 kΩ, but dropped down to 5 kΩ after this treatment. TGA showed that above 180 °C graphene and grapheneoxide show a strong mass loss. It is known that CO and CO$_2$ groups leave grapheneoxide if heated to this temperature. The mass loss and the increase of conductivity of graphene upon heating is probably due to an

Results and Discussion

incomplete reduction by hydrazine or because of the release of thermolabile nitrogen groups attached to graphene during hydrazine reduction. This proposal is supported by the fact that grapheneoxide reduced by hydrazine for a much longer periode (24 h) shows a reduced mass loss during heating.[22] However such long reduction times lead to an aggregation of the graphene flakes in solution.

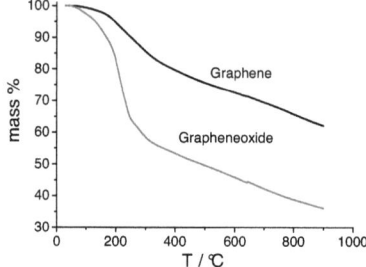

Figure 58. TGA of graphene and graphenoxide.

To obain more information about the chemical composition of the graphene, Auger analysis of a sample obtained by drying a sample droplet on a platinum surface was preformed. Only a small content of 3.4 atomic percent of oxygen and 2.4 atomic percent of nitrogen can be found in graphene obtained by reduction of graphene oxide (Fig. 59). Slightly higher values of these impurities have been found in literature by solvothermal reduction of grapheneoxide.[18] This proves the efficient removing of oxygen functionalities by the hydrazine reduction procedure.

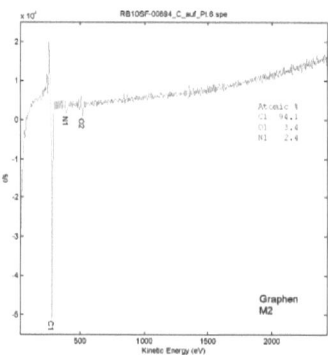

Figure 59. Auger spectra of graphene.

Results and Discussion

To characterize the conductance of a thin film graphene, 2 µl of the graphene suspension were dropcoated on an interdigitated electrode. Fig. 60 shows the influence of temperature on the conductance of the graphene film. Upon heating from 22°C to 85°C the resistance of the film is reduced by 65 %. Therefore the graphene film exhibits semiconducting properties, e.g. the resistance decreased with increasing temperature. A decrease in resistance during a temperature increase was found earlier.[23][24] A variable range hopping (VRH) charge-transport mechanism was suggested for reduced grapheneoxide. VRH involves consecutive inelastic tunneling processes between localized states and for a two dimensional system the conductivity is proportional to $T^{1/3}$. Hopping probably occurs between graphene regions which are separated by defects.[24]

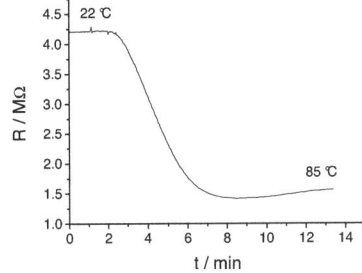

Figure 60. Influence of a temperature increase from 22°C to 85 °C on the resistance of a graphene film.

In-situ resistance measurements are commonly used to obtain information about the influence of oxidations or reductions on the resistance of e.g. conducting polymers. In the case of graphene such measurements were performed for epitaxial grown graphene and mechanical exfoliated graphene,[25],[26] but not for graphene obtained by reduction of graphene oxide. Fig. 61 shows the potential influence on the resistance of graphene on an interdigitated electrode. Going from 0.1 V towards more negative or positive potentials the resistance decreased, but for potentials more positiv than 0.6 V increased again. Excluding the increase of resistance at potentials more positive than 0.6 V the behaviour is similar to that reported for epitaxial grown and mechanical exfoliated graphene.[25],[26] The behaviour in this potential region can be explained by inducing both n- and p-carriers by capacitive charging of the polarizable graphene/electrolyte interface.[26] A decrease in resistance due to n- and

Results and Discussion

p-doping is also found in field effect transistors made from reduced graphene oxide[27]. At potential more positive than 0.6 V probably the oxidation of the graphene flakes starts, which causes the increase in resistance.

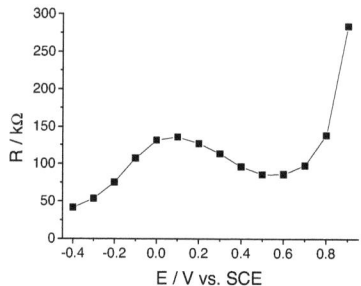

Figure 61. Potential dependent resistance of graphene in pH 7 buffer (10 mM) containing 0.1 M NaCl

3.7.2. Evaluation of graphene as sensor material for NO_2 sensing

The response of graphene towards nitrogen dioxide has been investigated in several works, due to the strong response of graphene upon NO_2 exposure.[1],[4]-[6],[8] Nevertheless no work investigated e.g. the effect of humidity on the response and regeneration of such sensors. An investigation of this influence of humidity and of the temperature influence on the response of graphene towards NO_2 and on its regeneration is presented in this chapter.

3.7.2.1. Results and Discussion

Humidity is known to influence the resistance of graphene. To investigate this influence the dry synthetic was humidified by passing it through the headspace of a water containing flask. Upon switching from dry air to humidified air the resistance droped by 19 % (Fig. 62 A). Furthermore the presence of humidity strongly affects the response of graphene towards NO_2 (Fig. 62). In presence of humidity a stronger and faster response towards NO_2 was observed. Additionally regeneration after NO_2

Results and Discussion

exposure was much faster compared to dry air. Similar effects were observed earlier for carbon nanotubes.[28]

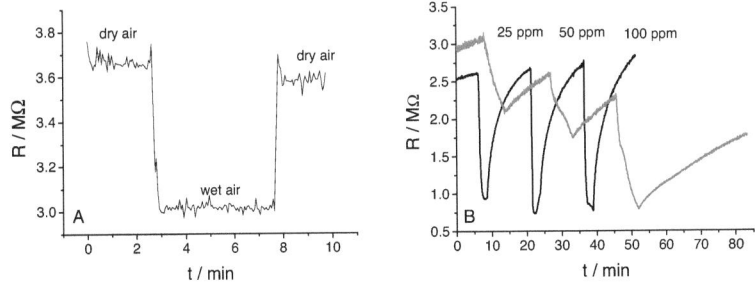

Figure 62. Influence of humidity on the resistance of graphene (A) and its response towards NO_2 (B).

Another parameter which influences the sensor behaviour is temperature. Due to the highly temperature dependent conductance of graphene (Fig. 60), measurements at a controlled temperature are much more reproducible. Heating of the sensor is a convenient technique to achieve a constant temperature. Furthermore heating is known to accelerate the regeneration of most materials used in gas sensors.[29] In Fig. 63 the normalized response of a graphene coated sensor towards 3 additions of 50 ppm NO_2 at room temperature is compared to the response at 85°C. One can see that already heating to moderate temperatures drastically accelerates the regeneration of the sensor, whereas the response time and magnitude was not affected. This is in contrast to the results reported in [4], where a decrease in sensitivity at higher temperatures was reported. Another positive effect of measuring at higher temperatures is the diminishing influence of humidity. Almost no difference was found in the response towards NO_2 using dry or humidified air. The operation temperature of 85 °C is still much lower than temperatures commonly used in metal oxide based gas sensors, which are usually operated at temperatures of several hundred degrees Celsius.[29]

Results and Discussion

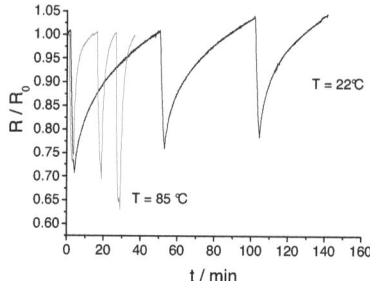

Figure 63. Influence of a temperature increase on normalized response to 50 ppm NO_2 and the regeneration.

In conlusion using graphene based gas sensors at elevated temperatures allows to diminish effects of temperature and humidity and ensures a fast sensor regeneration.

3.7.3. Graphene palladium nanoparticle layer by layer composite

One methode to induce selectivity and increase the sensitivity of graphene towards certain gases is to modify graphene with metallic nanoparticles.[2] Graphene can be decorated with metallic nanoparticles by first adding metal salt precursors to a solution of the graphene oxide, and then converting them to metallic nanoparticles by reduction.[30]-[33] Other methods include electrochemical deposition of metallic nanoparticles on thin layers of graphene.[9] Composites of graphene with gold,[33] ruthenium,[30] platinum[30],[32],[33] and palladium[30],[31],[33] nanoparticles were synthesized by this method. Composites of palladium nanoparticles with carbon nanotubes or graphene are known to be sensitive towards hydrogen[9],[34]-[35].

Layer-by-layer deposition (LbL), was shown to work also with graphene/polymer systems.[12] In this chapter the LbL deposition is used to obtain a composite material consisting of graphene and palladium nanoparticles, which is sensitive towards hydrogen.

Results and Discussion

3.7.3.1. Results and discussion

Graphene produced by the reduction of grapheneoxide is negatively charged.[12] To enable an adsorption of graphene, glass surfaces were modified by a quaterny ammonium silane. Furthermore it is well known that metallic nanoparticels adsorb at both positively[36] and negatively[37] charged surfaces. Therefore a layer by layer adsorption of graphene and palladium nanoparticles (PdNP) should be possible. This was confirmed by measuring the UV-Vis absorption after 2, 4 and 6 deposition cycles (Fig. 64). A gradual non-linear increase in the absorption over the whole spectral range was observed.

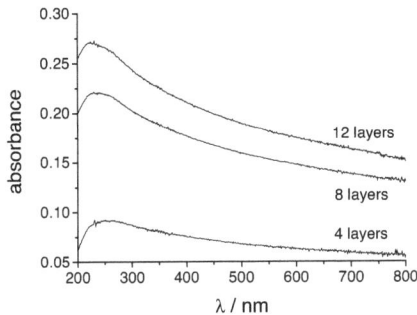

Figure 64. Absorption spectra of composite materials consisting of 4, 8 and 12 alternating layers of graphene and palladium nanoparticles.

LbL adsorption of graphene and palladium nanoparticles on gold was further studied by surface Plasmon resonance technique. Fig. 65 shows the change of the SPR signal during graphene and PdNP adsorption and the change of the SPR angle dependence from a clean surface to a 3 bilayer coated surface. One can see that the reflectance of the gold surface decreased during graphene adsorption. Due to the very thin layer of graphene the changes are relative small, however clear visible. During the adsorption of PdNP a much stronger increase of the reflectance occurred. The adsorption of the palladium nanoparticles starts to saturate after around 10 minutes. It has to be stated that after the following graphene adsorption, the adsorption kinetic of palladium nanoparticles was almost the same as after the first graphene layer. This confirms the adsorption of graphene on the PdNP. In Fig. 65 B

Results and Discussion

the angle dependence of the surface Plasmon resonance is shown for clean gold and after the first and second palladium adsorption step. The adsorption of palladium nanoparticles resulted in a dampening and a shift to higher angles of the Plasmon resonance accompanied by a peak broadening.

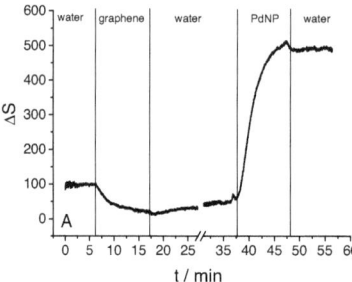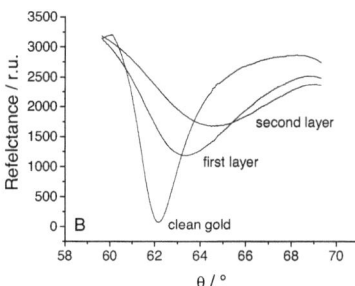

Figure 65. Change of the SPR signal at constant angle during the layer by layer deposition (a) and change in the shape of the angle dependence of the surface plasmon resonance due to assembling of the composite material (b).

SEM images of a 10 layer composite of graphene and PdNP are presented in Fig. 66. One can see a homogenous distribution of PdNP clusters as bright spots on the rough surface of the composite. No individual particles can be seen due to their small size (about 5 – 10 nm) (see Fig. 38). The size of the bright spots is in accordance with the data from light scattering, where a size of 20 nm to 50 nm was found. Some bigger agglomerates of PdNP can be also found on the surface.

Figure 66. SEM pictures of the graphene/PdNP composite (10 bilayers) on an interdigitated gold electrode taken in the gap between the gold stripes. The edge of the gold electrode can be seen on the top of the left picture.

Results and Discussion

To further proof the presence of Palladium on the electrode and evaluate its electrochemical activity, cyclic voltammograms in 0.5 M H_2SO_4 were recorded. Fig. 67 shows the voltammogram of an electrode coated by 10 bilayers. One can clearly see the palladium features, e.g. the strong hydrogen reduction reduction currents, the hydrogen desorption peak at 66 mV and the palladiumoxide reduction peak at 460 mV. The strong oxidation peak at 1.24 V corresponds to the oxidation of the gold electrode. The observation of the high hydrogen reduction currents shows, that the palladium-nanoparticles in the composite are catalytically active.

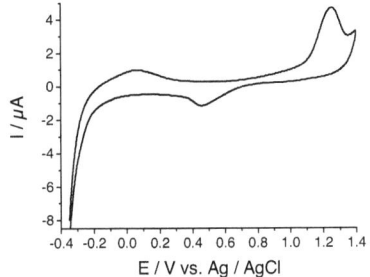

Figure 67. Cyclic voltammogram of a graphene/PdNP composite (10 bilayers) on a interdigitated gold electrode (in 0.5 M sulfuric acid). Scan rate: 0.1 V/s

In-situ resistance measurements of the composite in a pH 7 buffer show a strong decrease of the composite resistance upon n- or p-doping (Fig. 68). The resistance maximum and therefore the n- to p- transition is at 0.1 V vs. Ag / AgCl. This is similar to the potential dependent resistance for graphene described in chapter 3.7.1 (Fig. 61). A similar shape was also found for back gated field effect transistors using chemically derived graphene layers on SiO_2 electrochemically modified by Palladium nanoparticles[9] and liquid electrolyte gated graphene.[26] The possibility to increase the conductance of the composite by n- and p-doping should result in an increase in conductance upon exposure of the composite to both electron donating and electron accepting analytes.

Results and Discussion

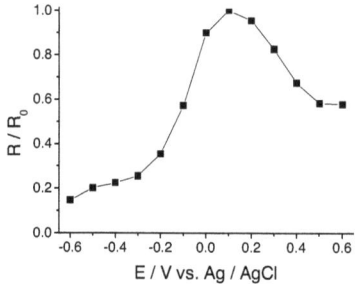

Figure 68. Normalized potential-dependent resistance of the graphene/PdNP composite (10 bilayers) in phosphate buffer (10mM) containing 0.1 M NaCl at pH 7.

In order to test the response of the graphene/PdNP composite towards hydrogen the sensor was placed in a flow cell and exposed to hydrogen diluted with synthetic air. A strong drop in the resistance of the sensors was observed upon exposure of two sensors coated by graphene/PdNP composites to hydrogen (Fig. 69). This is in accordance with the results from in-situ resistance measurements (Fig. 68) and corresponds to the data presented for palladium-modified graphene,[9] but is opposite to the data obtained with carbon nanotube/PdNP or PtNP and graphene/PtNP composites.[34]-[35] The reason for the resistance decrease is probably the dissolution and dissociation of hydrogen at the PdNPs. This may lead to a decrease of the particle's work function, resulting in a transfer of electrons to the graphene.[9] As expected, the composite with seven layers displayed a stronger response to hydrogen than the one with ten layers. Graphene without palladium nanoparticles displayed only a very small response to hydrogen. This clearly indicates the catalytic role of palladium nanoparticles in the n-doping of graphene by molecular hydrogen.

In humid air the regeneration of the sensor was complete within the time scale of a few minutes. In dry air the sensor response was almost irreversible. Similar effects were reported earlier.[9]

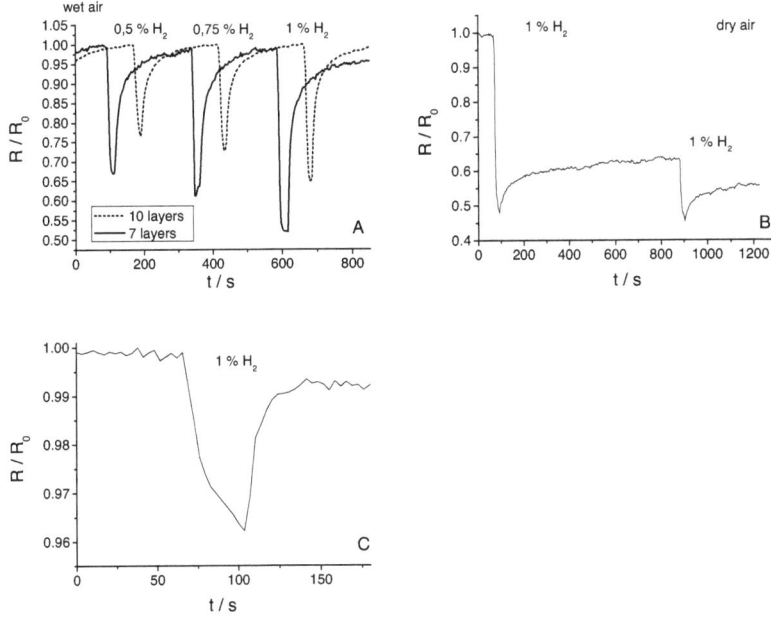

Figure 69. Response of graphene (c) and graphene/PdNP composites (a, b) with 7 or 10 layers to hydrogen diluted in wet (a, c) air and dry (b) air.

The influence of humidity on the resistance of graphene and the graphene palladium nanoparticle composite is show in Fig. 70. In both materials the resistance decreased by switching from humid to dry air, however the response was faster and higher for the graphene-palladium nanoparticle composite. In contrary the graphene covered sensor was more sensitive towards NO_2 than the graphene-palladium nanoparticle composite coated sensor (Fig. 70). This is probably due to the much thinner layer of graphene on this sensor, than on the graphene-palladium nanoparticle composite coated sensor.

Results and Discussion

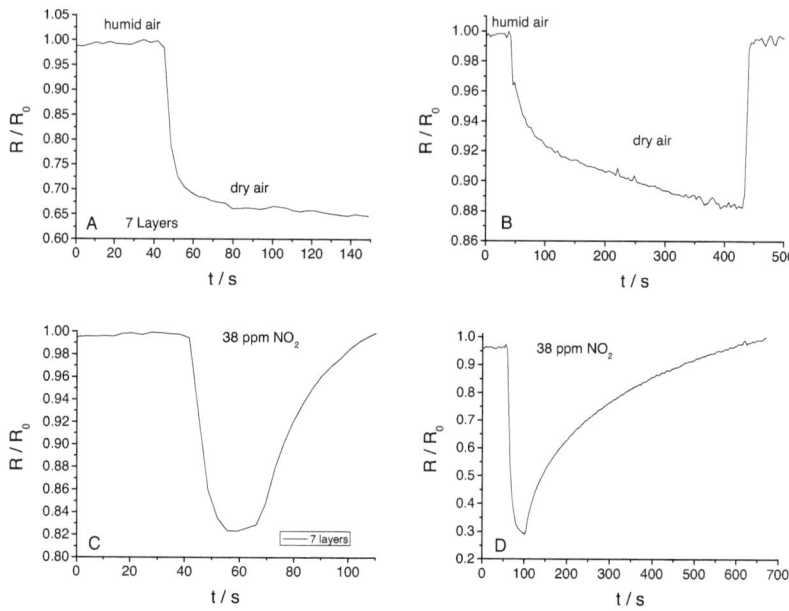

Figure 70. Response of the graphene/PdNP composite (7 layers) (a,c) and graphene (b,d) towards humid air and NO2.

Fig. 71 summarizes the responses of the two sensors on the exposure to hydrogen, NO_2 and humidity. Conceivable, one can distinguish between H_2, NO_2 and humidity by combining the two sensors. In fact, an array of different graphene-based composite materials should pave an efficient way for the selective analysis of gases.

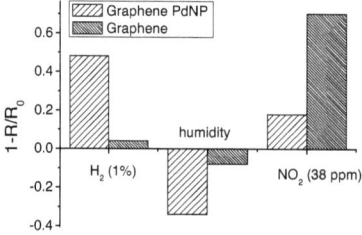

Figure 71. Comparison of the response of the graphene/PdNP composite and graphene to hydrogen, NO_2 and humidity.

3.7.4. Electrochemical modification of graphene with nanoparticles

Another possibility to modify graphene flakes by metallic nanoparticles is the electrochemical deposition of nanoparticles on preimmobilized graphene layers. Nanocrystalline nickel particles can be electrochemically deposited on graphite by an "H_2-coevolution method" using short reduction pulses at highly negative potentials.[38] As carbon monoxide is known to adsorb at nickel surfaces, resulting in changes of the nickel work function,[39] one can expect changes in the response to carbon monoxide upon modification of graphene by nickel nanoparticles. In this chapter the electrochemical modification of spin coated graphene layers on interdigitated gold electrodes by nickel nanoparticles is described and the resulting changes of the resistance response to carbon monoxide are evaluated.

3.7.4.1. Results and Discussion

Graphene layers were spin-coated on interdigitated gold electrodes by dropping 10 µL of a solution of graphene in water (~0.5 g / l) diluted 1:1 by ethanol on the sensor chip and spinning off the solution using a home made spin coating system. To modify the resulting electrodes by nickel nanoparticles, nickel reduction was carried out from a solution of nickel sulfate (10 mM) adjusted to pH 8.3, by applying a short reduction pulse (0.5 s) at -1.52 V vs. Ag / AgCl. This leads to the formation of nickel particles with a size of about 20 – 100 nm on the graphene flakes, which are contacted by the gold electrodes (Fig. 72). Of course a much higher density of nickel nanoparticels was achieved on the gold contacts due to their much higher conductivity. To achieve a better coverage of the whole gap with nickel nanoparticle modified graphene, more homogeneous coatings of graphene have to be achieved.

Results and Discussion

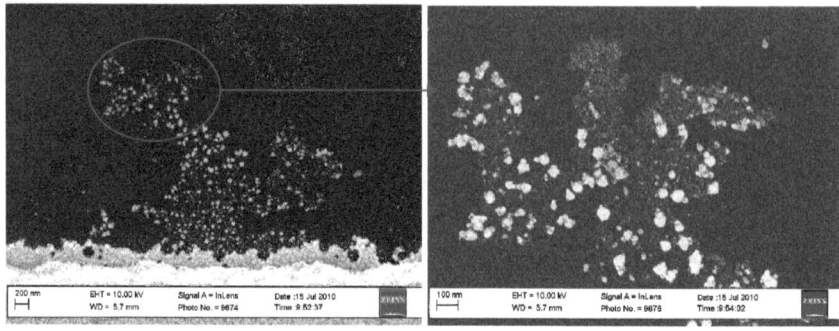

Figure 72. SEM images of nickel nanoparticel modified graphene on an interdigitated gold electrode at two different magnifications. The gold electrode also covered by nickel nanoparticles can be seen on the bottom of the left image

Despite the incomplete coverage of the electrode by nickel nanoparticles a difference in the response to carbon monoxide was found at 85° C. Interestingly the resistance of graphene increased upon exposure to 150 ppm carbon monoxide at room temperature (dashed line in Fig. 73), however showed almost no change on exposure to 150 ppm carbon monoxide at 85 °C (grey line in Fig. 73). In contrast with the nickel nanoparticle coated electrode the resistance increase upon exposure to 150 ppm carbon monoxide was also observed at 85° C (black line in Fig. 73). Probably adsorption of carbon monoxide to graphene is only weak and takes not place at higher temperatures, but is enhanced by coating the graphene with nickel nanoparticels. The response to 150 ppm CO was weak (~ 4%), but clearly visible. A stronger response may be possible by optimisation of the graphene coating.

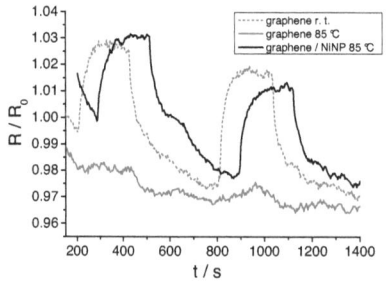

Figure 73. Response of graphene and nickel nanoparticle modified graphene on interdigitated gold electrodes upon exposure to 150 ppm carbon dioxide.

Results and Discussion

It was shown in this chapter that electrochemcial modification of graphene by metallic nanoparticles provides a simple and efficient way to tune the sensitivity and therefore also the selectivity towards various gaseous analytes. A combination of different modified sensors into an array can lead to an effective tool for the quantification of various gaseous analytes.

3.7.5. Experimental

Graphene synthesis and characterisation

Graphene oxide was synthesized using a modified Hummers method.[10],[11] In detail, 50 mg of graphite (Thielmann, flake graphite) were added to a mixture of 37.5 mg $NaNO_3$, 3.75 ml of conc. H_2SO_4 and 225 mg $KMnO_4$. The mixture was sonicated for 3 h and then stirred at room temperature for 3 days. Subsequently, 7 mL of 5% H_2SO_4 were added and the solution was stirred at 100 °C for 2 h. After addition of 150 µL of 30% H_2O_2 a yellow suspension was obtained. The mixture was centrifuged for 5 min and the supernatant removed. The product was re-suspended in 3% H_2SO_4 containing 0.5% of H_2O_2 and again centrifuged for 5 min. This step was repeated four times, another two times with 3% HCl, and finally two times with deionized water. As the pH increased by washing with deionized water, a stable suspension of graphene oxide was obtained. To further eliminate impurities, the suspension was dialysed against deionized water. To obtain chemically derived graphene, 7 ml of a suspension of the graphene oxide (0.5 g / l) were added to 31 µl of 32% NH_3 and 2.5 µl hydrazine hydrate. Reduction was carried out by stirring the mixture at 100 °C for 1 h.[12] The resulting stable black suspension was diluted 5 times with water for further use.

Electrodes were prepared by dropcoating 1 µl of the graphene suspension (0.01 g / l) onto the electrode head on a hot plate.

Raman and SEM samples on SiO_2 and platinum surfaces were obtained by the same procedure.

SEM images were obtained on a Zeiss SEM-device. Measurement parameters are given in the corresponding figures.

Raman spectra were acquired using a micro-Raman imagine device with a 488 nm Ar-ion Laser.

Results and Discussion

Auger-analysis was done by Infineon AG.

Layer by layer deposition of graphene and palladium nanoparticels

Palladium nanoparticles were synthesized as desribed in chapter 3.5.3.

Glass surfaces were first modified with an adhesive layer of quaterny ammonium silane by dipping the glass slides for 12 h into a solution of 20 μL N-Trimethoxysilylpropyl-N,N,N-trimethylammonium chloride in ethanol. Then the slides were washed with ethanol and used for LbL deposition. Gold surfaces were used for layer by layer deposition of graphene and palladium nanoparticles without adhesive layer. Graphene suspensions with a concentration of 0.1 g / l were used for LbL deposition. The adsorption of graphene and palladium nanoparticles on gold surfaces was confirmed by surface plasmon resonance measurements (Biosuplar).

LbL deposition was carried out on modified quartz glass for UV-VIS spectroscopy and on linear gold electrodes for conductivity measurements. Resistance was measured as described in chapter 2.

Electrochemical modification of graphene

Graphene layers were spin-coated on interdigitated gold electrodes by dropping 10 μL of a solution of graphene in water (~0.5 g / l) diluted 1:1 by ethanol on the sensor chip and spinning off the solution using a home made spin coating system. Electrochemical deposition of nickel nanoparticels was carried out from a solution of 10 mM $NiSO_4$ containing 1 M NaCl and 1 M NH_3 adjusted to pH 8.3, by HCl by applying a 0.5 s reduction pulse at -1.52 V vs. Ag / AgCl using a CHI-660A electrochemical analyzer. A platinum counter electrode was used.

Gas measurements

Gas measurements were carried out using a home-made computer controlled gas mixing system. 1% hydrogen, 300 ppm CO or 300 ppm NO_2 in synthetic air were diluted to the corresponding concentration with synthetic air. If necessary the gas was then piped through the headspace of a water containing flask in order to humidify it. A temperature controlled flow cell was used in all measurements.

3.7.6. References

[1] F. Schedin, A. K. Geim, S. V. Morozov, E. W. Hill, P. Blake, M. I. Katsnelson, K. S. Novoselov, Nat. Mater. **2007**, 6, 652-655.

[2] K. R. Ratinac, W. Yang, S. P. Ringer, F. Braet, Environ. Sci. Technol. **2010**, 44, 1167-1176.
[3] Y. Dan, Y. Lu, N. J. Kybert, Z. Luo, A. T. C. Johnson, Nano Lett. **2009**, 9, 1472-1475.
[4] J. D. Fowler, M. J. Allen, V. C. Tung, Y. Yang, R. B. Kaner, B. H. Weiller, ACS Nano **2009**, 3, 301-306.
[5] G. Lu, L. Ocola, J. Chen, Nanotechnol. **2009**, 20, 445502.
[6] G. Lu, L. E. Ocola, J. Chen, Appl. Phys. Lett. **2009**, 94, 083111-3.
[7] J. T. Robinson, F. K. Perkins, E. S. Snow, Z. Wei, P. E. Sheehan, Nano Lett. **2008**, 8, 3137-3140.
[8] V. Dua, S. Surwade, S. Ammu, S. Agnihotra, S. Jain, K. Roberts, S. Park, R. Ruoff, S. Manohar, Angew. Chem. **2010**, 122, 2200-2203.
[9] R. S. Sundaram, C. Gomez-Navarro, K. Balasubramanian, M. Burghard, K. Kern, Adv.Mater. **2008**, 20, 3050-3053.
[10] W. S. Hummers, R. E. Offeman, J. Am. Chem. Soc. **1958**, 80, 1339.
[11] H. Becerril, J. Mao, Z. Liu, R. Stoltenberg, Z. Bao, Y. Chen, ACS Nano **2008**, 2, 463-470.
[12] D. Li, M. B. Muller, S. Gilje, R. B. Kaner, G. G. Wallace, Nat. Nano **2008**, 3, 101-105.
[13] A. C. Ferrari, J. Robertson, Phys. Rev. B **2000**, 61, 14095.
[14] Y. Kawashima, G. Katagiri, Phys. Rev. B **1995**, 52, 10053–10059.
[15] F. Tuinstra, J. L. Koenig, J. Chem. Phys. **1970**, 53, 1126-1130.
[16] V. López, R. Sundaram, C. Gómez-Navarro, D. Olea, M. Burghard, J. Gómez-Herrero, F. Zamora, K. Kern, Adv. Mater. **2009**, 21, 4683-4686.
[17] W. Gao, L. Alemany, L. Ci, P. Ajayan, Nat. Chem. **2009**, 1, 403-408.
[18] H. Wang, J. T. Robinson, X. Li, H. Dai, J. Am. Chem. Soc. **2009**, 131, 9910-9911.
[19] D. Yang, A. Velamakanni, G. Bozoklu, S. Park, M. Stoller, R. D. Piner, S. Stankovich, I. Jung, D. A. Field, C. A. Ventrice Jr., R. S. Ruoff, Carbon **2009**, 47, 145-152.
[20] O. Akhavan, Carbon **2010**, 48, 509-519.
[21] D. R. Dreyer, S. Park, C. W. Bielawski, R. S. Ruoff, Chem. Soc. Rev. **2010**, 39, 228-240.
[22] S. Stankovich, D. A. Dikin, R. D. Piner, K. A. Kohlhaas, A. Kleinhammes, Y. Jia,

Y. Wu, S. T. Nguyen, R. S. Ruoff, Carbon **2007**, 45, 1558-1565.
[23] S. Gilje, S. Han, M. Wang, K. L. Wang, R. B. Kaner, Nano Lett. **2007**, 7, 3394-3398.
[24] C. Gómez-Navarro, R. T. Weitz, A. M. Bittner, M. Scolari, A. Mews, M. Burghard, K. Kern, Nano Lett. **2007**, 7, 3499-3503.
[25] Y. Ohno, K. Maehashi, Y. Yamashiro, K. Matsumoto, Nano Lett. **2009**, 9, 3318-3322.
[26] P. K. Ang, W. Chen, A. T. S. Wee, K. P. Loh, J. Am. Chem. Soc. **2008**, 130, 14392-14393.
[27] G. Eda, G. Fanchini, M. Chhowalla, Nat. Nano **2008**, 3, 270-274.
[28] C. Cantalini, L. Valentini, I. Armentano, L. Lozzi, J. M. Kenny, S. Santucci, Sens. Actuators B **2003**, 95, 195-202.
[29] D. Kauffman, A. Star, Angew. Chem. Intern. Ed. **2008**, 47, 6550-6570.
[30] K. Gotoh, K. Kawabata, E. Fujii, K. Morishige, T. Kinumoto, Y. Miyazaki, H. Ishida, Carbon **2009**, 47, 2120-2124.
[31] Z. Hu, M. Aizawa, Z. Wang, N. Yoshizawa, H. Hatori, Langmuir **2010**, DOI doi: 10.1021/la9040166.
[32] Y. Si, E. T. Samulski, Chem.Mater. **2008**, 20, 6792-6797.
[33] C. Xu, X. Wang, J. Zhu, J.Phys.Chem.C **2008**, 112, 19841-19845.
[34] S. Mubeen, T. Zhang, B. Yoo, M. A. Deshusses, N. V. Myung, J. Phys. Chem. C **2007**, 111, 6321-6327.
[35] A. Kaniyoor, R. Jafri, T. Arockiadoss, S. Ramaprabhu, Nanoscale **2009**, 1, 382-386.
[36] U. Lange, S. Ivanov, V. Lyutov, V. Tsakova, V. Mirsky, J. Solid State Electrochem. **2010**, 14, 1261-1268.
[37] F. Kurniawan, V. Tsakova, V. Mirsky, Electroanal. **2006**, 18, 1937-1942.
[38] M. P. Zach, R. M. Penner, Adv. Mater. **2000**, 12, 878-883.
[39] J. Bertolini, G. Dalmai-Imelik, J. Rousseau, Surface Sci. **1977**, 68, 539-546.

4. Conclusion

Simultaneous two- and four-point measurements are used to characterize the polymer bulk as well as the contact resistance of polypyrrole and polythiophene films. Polythiophene films obtained were electrochemically and spectroelectrochemically characterized in aqueous and organic electrolytes.

Sensor chips containing not only four electrodes for resistance measurements, but additionally two electrodes which can be used as reference and counter electrodes, where used for electrochemical transistor design. Electrochemical transistors based on polyaniline were characterized in aqueous electrolytes. Additionally a polythiophene based electrochemical transistor was made using a gel electrolyte instead of an aqueous electrolyte. This quasi-solid state electrochemical transistor was used as gas sensor, allowing an electrochemical regeneration of the sensor after analyte exposure.

In electrochemical transistors not only the conducting film between the electrodes for resistance measurements, but also the gate (reference) electrode can be used as chemosensitive element. This was demonstrated by using an ion-selective electrode as gate electrode. A shift of the potential of the ion-selective electrode leads to changes in the resistance of the conducting polymer film between the electrodes for resistance measurements.

Layer-by-layer technique was shown to be an effective methode for fabrication of conducting polymer / metal nanoparticle composite materials. Immobilisation of gold- and palladium nanoparticles in a polyaniline matrix was demonstrated. It was shown that the immobilized nanoparticles can work as receptors for specific analytes. Additionally the electrocatalytic properties of the immobilized palladium nanoparticels towards hydrazine oxidation were demonstrated.

Conducting polymer / metal nanoparticle composites were also made by a dispersion of palladium nanoparticels in a PEDOT PSS solution. This dispersion has the advantage that it can be easily coated on almost any substrate. Films obtained from this solution were characterized and used in a chemoresistor for hydrazine and NADH.

Graphene is a new promising material for applications in chemical sensors. Graphene made by reduction of graphene oxide was characterized and its application in gas sensors was evaluated. Furthermore it was shown that

Conclusion

modification of graphene by metal nanoparticles can change its sensitivity towards gaseous analytes. Combination of different modified sensors in an array should lead to an improved selectivity of such sensors.

Acknowledgements

I want to express my sincere gratitude to the following people who contributed to the success of this work.

First of all, I am very grateful to **Prof Vladimir M. Mirsky** for offering me the possibility to do this work under his supervision, for his extensive help and many fruitful discussions about this work.
I am also very grateful to **Prof. Otto S. Wolfbeis** for giving me the opportunity to do this work at the Institut of Analytical Chemistry, Chemo- and Biosensors and for his help and support during my work.

I would like to say special thanks to **Dr. Vessela Tsakova** for the organisation of two research visits at the Institute of Physical Chemistry at the Bulgarian Academy of Science in Sofia, for planning the work with the polyaniline / metal nanoparticle composites and the fruitful discussion about the results obtained in that DAAD project. Special thanks also to **Dr. Svetlozar Ivanov** and **Vladimir Lyutov** for the always interesting and very successful cooperation in the frame of that project.

I have to thank **Prof. Michael Vorotyntsev and Dr. D. Konev** for the organisation of two research visits at the University of Bourgogne in Dijon and the planning and discussions of the work I did there.

Prof. Werner Kunz for financial support in the purchase of new electrodes and the cooperation in the sensor development for caproic and lactic acid. Thank you also to Jeremy Drapeau for his cooperation in this project.

I am likewise thankful to the following people for the help and support of this work:

Dr. Günter Ruhl from Infineon AG for the installation of a cooperation between Infineon and our Institut in the field of graphene based gas sensors, for heat treatement experiments of graphene samples, for SEM measurements and for the Auger analysis of graphene.

Dr. Jonathan Eroms and **Cornelia Linz** from the Institut of Experimental and Applied Physics at the University of Regensburg for SEM measurments.

Jürgen Kolouch and **Peter Hausler** from the University of Applied Science in Regensburg for Raman measurements.

Dr. Rainer Müller for TGA-measurements.

All people at the Institut of Physical chemistry at the Bulgarian Academy of Science who were involved in the TEM and SEM measurements.

I am also very grateful to all my colleagues at the Institut of Analytical chemistry, Chemo- and Biosensors, especially **Dr. Thomas Hirsch** and **Dr. Natalyia Roznyatovskaya** for many fruitful discussions and help with measurments.
I also have to thank **Angela Haberkern**, **Rosemarie Walter**, **Joachim Rewitzer** and **Angela Stoiber** for their technical assistance and **Edeltraut Schmid** and **Sabine Rudloff** for the friendly assistance in any official or personal business.

Last but not least I am very grateful to my parents **Rita & Volkmar Lange** for their moral and financial support during my studies.

I want morebooks!

Buy your books fast and straightforward online - at one of world's fastest growing online book stores! Environmentally sound due to Print-on-Demand technologies.

Buy your books online at
www.morebooks.shop

Kaufen Sie Ihre Bücher schnell und unkompliziert online – auf einer der am schnellsten wachsenden Buchhandelsplattformen weltweit! Dank Print-On-Demand umwelt- und ressourcenschonend produziert.

Bücher schneller online kaufen
www.morebooks.shop

KS OmniScriptum Publishing
Brivibas gatve 197
LV-1039 Riga, Latvia
Telefax: +371 686 204 55

info@omniscriptum.com
www.omniscriptum.com

Printed by Books on Demand GmbH, Norderstedt / Germany